Know Your Oscilloscope

by

Robert G. Middleton

Howard W. Sams & Co., Inc.
4300 WEST 62ND ST. INDIANAPOLIS, INDIANA 46268 USA

Preface

Since the previous edition of this book was written, oscilloscopes have become more sophisticated and versatile. Today, many service-type oscilloscopes have dual-channel design. Triggered sweeps are provided in most scopes, and amplifier bandwidths have been considerably increased. For example, a modern dual-trace triggered-sweep 20-MHz oscilloscope is illustrated in Fig. 1-14. Although this would have been regarded as a lab-type oscilloscope in former years, it now falls in the category of service-type instruments.

With the advent of the digital revolution, specialized high-technology oscilloscopes, such as the one shown in Fig. 11-20, tend to dominate the lab-type category of designer's and troubleshooter's display instruments. A digital data-field display is distinguished from a conventional analog (time/frequency) waveform display, in that the former shows sequences of digital events as 1s and 0s with respect to computer clock time, whereas, the latter shows the variation of an electrical quantity with real time (the actual time in which physical events occur). However, it is essential to *start at the beginning,* and Chapter 1 starts with a review of the simplest possible oscilloscope.

After establishing the basic principles of a cathode-ray screen display, the text continues with a description of simple time bases, the essentials of waveform amplification, time-base synchronization, sweep triggering and expansion, and the fundamentals of dual-trace operation.

Specialized power supplies for oscilloscopes are the subject of the second chapter, with notes on the extra-high voltage requirements of scopes that have extremely fast writing speeds, such as those used in digital troubleshooting procedures. Power-supply regulation is explained, and the chief types of power-supply circuitry are reviewed.

Chapter 3 is concerned with sweep systems that are used in present-day oscilloscopes. Attention is given to the operation of triggered-sweep time bases, whereby, a small interval of a waveform can be selected and greatly expanded on the screen—for measurement of rise time or fall time, examination of waveform detail, identification of narrow transients, and so on. In Chapter 4, principles of synchronization are covered. The origin of "jitter" is pointed out; specific examples of the utility of external sync are cited, and triggering by "one-shot" transients is discussed.

Vertical-amplifier circuit action is treated in Chapter 5. The function of vertical-channel delay lines in the display of fast transients is explained. Vertical step-attenuator compensation is explained. Digital memories, in vertical-amplifier channels, that are for display of data fields are noted. Chapter 6 describes various kinds of oscilloscope probes, including the direct, isolating, low-capacitance, demodulator, capacitance-divider, and current-responding designs. Appropriate areas of probe application are defined and illustrated.

Supplementary oscilloscope equipment, including semiconductor quick checkers, audio oscillators, fm-stereo generators, square-wave generators, and high-frequency generators, are covered in Chapter 7. Audio-distortion, stereo-separation, music-power, alignment, and vectorgram tests are described and illustrated.

Chapter 8 is concerned with adjusting and servicing the oscilloscope. Procedures are explained for tracking down the source of pattern distortion, lack of normal brightness, out-of-calibration attenuators, and incorrect sweep speeds. Sensitivity measurement and frequency-response determination, transient (square-wave) response, and deflection linearity are discussed. The use of manufacturers' troubleshooting charts is exemplified. Then, frequency and phase measurements are detailed in the ninth chapter. The basic distinction between measurement applications and indicator applications is noted. The development of Lissajous figures is presented, with examples of their frequency and phase information.

Chapter 10 covers the important topic of amplifier testing. Frequency response, distortion, noise output, phase shifts, tone-control action, stability, equalization characteristics, and power-output determinations are described and illustrated. Evaluation of distortion products is discussed, and a comprehensive development of square-wave responses is provided. Common malfunctions are noted and the chapter concludes with a practical discussion of hum patterns. The final chapter gives details on the servicing of digital equipment, which is becoming increasingly important in modern electronic equipment.

In summary, this book is devoted to practical applications of the oscilloscope, in addition to its detailed examination of the inner workings of this versatile instrument. Although this text is primarily directed to the needs of the practical electronics technician, it will also be found of relevance to courses of instruction in technical institutes, vocational schools, and junior colleges. In order for the reader to obtain maximum value from the contents, I would stress the importance of actually working with the equipment as various procedures are described. This "reinforced learning," gained at the workbench, will prove to be much more effective than the knowledge that you can acquire from just reading the book.

ROBERT G. MIDDLETON

Contents

CHAPTER 1

Introduction

OSCILLOSCOPE BASICS

The word "oscilloscope" can be separated into two parts, "oscillo" and "scope." The first is short for "oscillations" and the second means "to view or see." Thus, if we take the word literally, it describes an instrument for viewing oscillations. Conventional oscilloscopes display electrical signals that vary with time, such as sine waves. When used with a suitable transducer, a conventional oscilloscope will provide a visual display of any physical quantity that can be reproduced as a voltage. For example, in a hospital, the heartbeat of a patient can be displayed on a crt screen. This is exemplified by the "biomedical test" waveform shown in Chart 1-1. With the advent of the digital revolution, specialized oscilloscopes have been made that provide displays of data fields, as exemplified in Fig. 1-1. A logic-state analyzer oscilloscope is shown in Fig. 1-2. The basic distinction between a digital data-field display and a sine-wave display is that the former presents sequences of digital events with respect to clock time, whereas the latter presents an analog variation of an electrical quantity with respect to real time. (The "clock" in a digital system is a crystal-controlled oscillator which synchronizes system operation; on the other hand, "real-time" has to do with the actual time in which physical events occur.)

A schematic for the simplest type of an oscilloscope is depicted in Fig. 1-3. It consists of a cathode-ray tube, a power supply, and two RC coupling circuits. The components are illustrated in Fig. 1-4. Although this elementary arrangement can be used in certain applications, such as modulation monitoring and for the display of vectorgrams in high-level chroma circuits, it has various disadvantages. These are:

1. A directly driven crt is comparatively insensitive; a typical crt requires approximately 300 volts peak-to-peak for full-screen

Chart 1-1. Oscilloscope

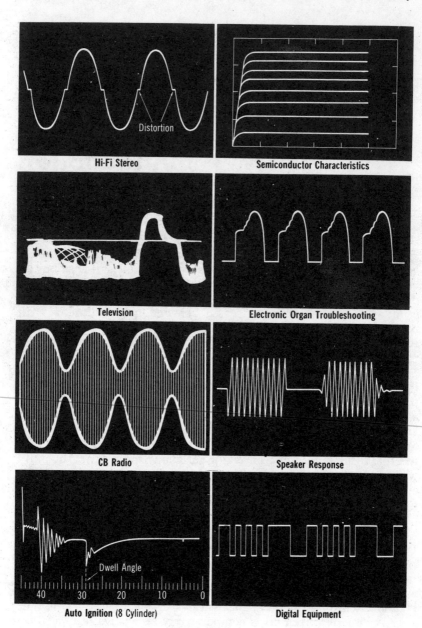

Hi-Fi Stereo

Semiconductor Characteristics

Television

Electronic Organ Troubleshooting

CB Radio

Speaker Response

Auto Ignition (8 Cylinder)

Digital Equipment

Fields of Application

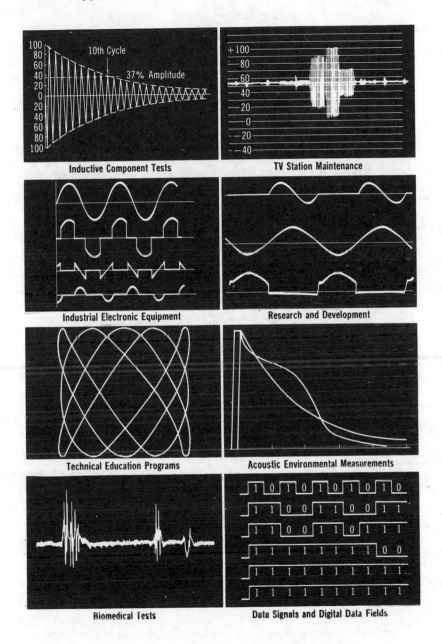

Inductive Component Tests

TV Station Maintenance

Industrial Electronic Equipment

Research and Development

Technical Education Programs

Acoustic Environmental Measurements

Biomedical Tests

Data Signals and Digital Data Fields

(A) Elementary display pattern.

```
0010  0010  1110  0010          11  0000
0010  0011  1110  1010          00  1111
0010  0100  1111  0100          10  1111
0010  0101  0001  1100          01  1111

0010  0110  0100  0011
0010  0111  1111  0000

0010  1000  1111  1010
0010  1001  1111  0101

0010  1010  1110  0010
0010  10f1  1110  1010

0010  1100  1111  0100
0010  1101  0001  1100

0010  1110  0100  0011
0010  1111  00.00  0000

0011  0000  1111  0000
0011  0001  1111  1010
```

(B) Check of a dual-clock operation displayed on screen of a data-domain analyzer.

Courtesy Hewlett-Packard

Fig. 1-1. Data-domain displays.

deflection. Therefore, vertical and horizontal amplifiers *are required* in the great majority of applications.

2. Single-ended drive to a crt results in variation of the focus (astigmatism) from one region of the screen to another. Accordingly, good focus *requires* push-pull (balanced or double-ended) drive to the crt deflection plates.

3. Unless an external sawtooth generator is used, the arrangement in Fig. 1-3 can display only Lissajous patterns and related waveforms. Because most applications utilize voltage-time displays, *it is desirable or necessary* to include a linear time base in the oscilloscope arrangement.

4. Although Lissajous figures are often self-synchronized, most voltage-time displays will not be synchronized (locked) on the crt screen unless the time base is automatically kept in step with the

Fig. 1-2. A logic-state analyzer oscilloscope used as a data-domain display.

displayed waveform. In turn, *it is desirable or necessary* to include a synchronizing network in the oscilloscope arrangement.

5. In many applications, it is helpful or necessary to move the pattern about on the crt screen, in order to adjust the height of the pattern, or to adjust the length of the pattern. Hence, a general-purpose oscilloscope *must include* vertical and horizon-

Fig. 1-3. The schematic for a simple type of oscilloscope.

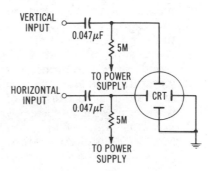

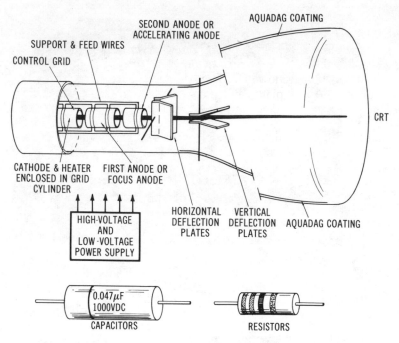

Fig. 1-4. The components needed for the circuit shown in Fig. 1-3.

tal-positioning controls, vertical-gain control(s), and horizontal-gain control(s).

6. Applications in state-of-the-art electronic circuitry often require that a selected interval of a waveform be "picked out" and expanded on the crt screen. Therefore, modern oscilloscopes *generally have* triggered sweeps.

7. Service data generally specify peak-to-peak voltage values of waveforms, and some applications require time measurements from one point on a waveform to another (for example, rise-time measurements are sometimes required). Accordingly, modern oscilloscopes *generally provide* calibrated vertical-step attenuators, and calibrated time bases.

8. Some applications are facilitated by simultaneous display of input/output waveforms, by simultaneous display of a complete waveform and, also, an expanded portion of a selected interval, or by simultaneous display of a reference waveform and related waveforms within a system. In turn, *it is sometimes desirable or necessary* to provide dual-trace display in an oscilloscope. Moreover, in digital-electronic applications, multichannel displays *are provided* by data-domain–type oscilloscopes.

CATHODE-RAY TUBES

Two basic types of cathode-ray tubes are used in the majority of oscilloscopes, as depicted in Fig. 1-5. The most common type is the single-beam tube, comprising an electron gun with a cathode, heater, and control grid, followed by a focusing anode, an accelerating anode, vertical-deflection plates, horizontal-deflection plates, and a fluorescent screen. The electron beam is electrostatically focused and deflected in this design. Lab-type oscilloscopes may employ a dual-beam tube, which includes two electron guns, two sets of anodes, and two sets of deflecting plates (Fig. 1-5B). Each basic type of crt has certain advantages in particular applications. When a dual-beam crt is used, two patterns can be displayed on the screen just as if two separate oscilloscopes were being employed. On the other hand, when a single-beam crt is used, two patterns must be time-shared in order to have a simultaneous display, as shown in Fig. 1-6. In alternate-channel operation, one complete waveform is traced on the screen; then, another complete waveform is traced; the first waveform is then completely retraced, and so on. However, in chopped operation (Fig. 1-6B), the two waveforms are rapidly sampled, and successive samples are displayed on the screen, as indicated.

High-frequency waveforms are displayed to their best advantage by alternate-channel operation in the dual-trace mode; conversely, low-frequency waveforms can be displayed with less flicker by chopped operation in the dual-trace mode. (Fig. 1-7 is a photograph of a highly sophisticated dual-trace oscilloscope.) When precise phase measurements are made between Channel A and Channel B displays, a single-beam crt provides better accuracy than a dual-beam crt. In addition, *persistence* is a factor that enters into consideration. As the crt beam moves across the face of the tube in response to voltages on the deflection plates, the corresponding spot of light also moves. If this movement is fast enough, the spot (or succession of dots) appears as a continuous trace or line of light. The blending of successive positions of the spot into an apparently continuous trace is due to two factors:

1. Persistence of the phosphor used in fabrication of the screen.
2. Persistence of vision.

Persistence of vision is the ability of the human eye to see any object or spot of light at its original position for a fraction of a second after it has moved. Similarly, the persistence of phosphor is its brief glow after the electron beam has left its initial spot.

In general-purpose oscilloscopes, the blending of the spot into a line is due almost entirely to the persistence of vision. In special-purpose oscilloscopes, a crt with a long-persistence phosphor may be used, and electrical phenomena of short duration and nonrepetitive form can be displayed. (Special storage-type oscilloscopes can retain a selected

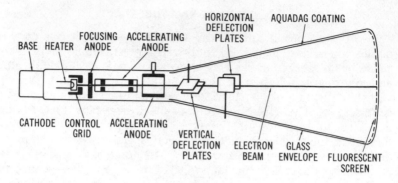

(A) Single-beam tube.

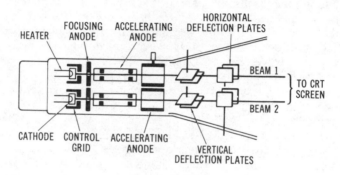

(B) Dual-beam tube.

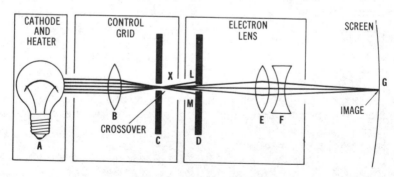

(C) Light analogy to the crt beam-focusing system.

Fig. 1-5. Two basic types of a crt.

14

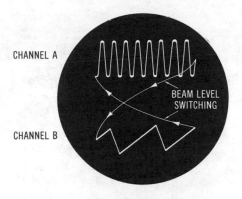

(A) Alternate channel operation.

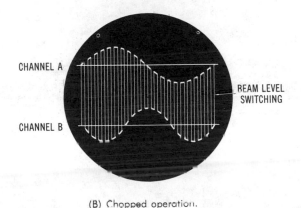

(B) Chopped operation.

Fig. 1-6. Dual-trace modes of operation.

pattern on-screen for an hour, or more; the stored pattern can be erased at any desired time.) The phosphor most commonly used in the crt of an oscilloscope is termed P1, and is rated for medium persistence. P5 is a phosphor of short persistence; P7 has a long persistence. Some phosphors, such as P4 (commonly used in tv picture tubes) and P11 (which provides a blue trace that is easily photographed) are widely used. Other types of phosphors produce green traces and red traces; these are utilized in color-tv picture tubes. Colored traces also find an application in special-purpose crt's for particular military operations.

One of the special crt's used in storage-type oscilloscopes uses a metal mesh construction behind the fluorescent screen, as shown in Fig. 1-8. The intensity of the electrostatic field in the apertures of the

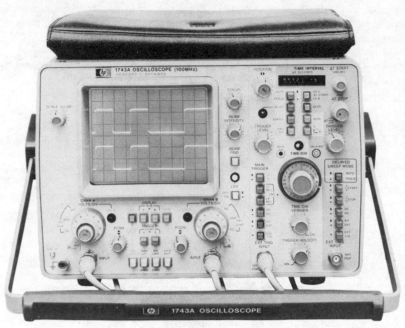

Fig. 1-7. A highly sophisticated dual-trace oscilloscope.

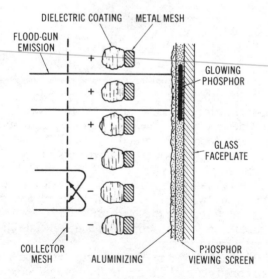

Fig. 1-8. The metal mesh design of a storage-type crt.

metal mesh determines whether the flood-gun electron emission will penetrate to the fluorescent screen. The metal mesh is usually operated at zero volts so that the flood-gun emission cannot go through to the fluorescent screen. However, the radiation from the trace that has been "written" on the screen modifies the electrostatic field intensity in the apertures of the metal mesh. The field intensity in the apertures that are behind the pattern on the screen has less repelling action on the electrons from the flood gun. In turn, the flood-gun electrons can pass through these apertures. The aluminized coating behind the phosphor viewing screen, in turn, accelerates the flood-gun electrons, and these electrons maintain the pattern that is glowing on the screen for an indefinite period. If the operator wishes to erase the pattern, he pushes a bias control that biases off the metal mesh so that no flood-gun electrons can pass. In turn, the glowing pattern disappears from the screen. Note that the flood-gun electrons that do not pass through the metal mesh are collected by the collector mesh.

Storage-type oscilloscopes provide improved displays in various applications. Even a brief nonrepetitive transient can be captured and "frozen" on the screen of a storage-type crt. The storage mode of operation is advantageous for observing changes in a signal, to compare the result of making a circuit adjustment, to compare the performance of two or more equipment setups, to compare an arbitrary waveform against a standard waveform, for reducing flicker in the display of waveforms with low repetition rates, for effectively displaying digital signals that have a low duty cycle, for reducing noise interference on a recurrent waveform, and for monitoring intermittent conditions.

GRAPH PATTERNS

To use a simple analogy, the electron beam can be considered as pencil writing upon the screen of the cathode-ray tube according to the voltage on the deflection plates. When a horizontal-deflection system is used (and practically no oscilloscope is built without one), the trace on the screen is really a graph. Graphs are now so commonplace that hardly a person has not seen one. Some examples are the temperature graphs and electrocardiographs used in hospitals, or the sales graphs of a business office.

The reader probably has drawn graphs in school; he will remember that they show two sets of data with the values of one set varying in some fashion as the other set varies. One set of values is plotted along the horizontal or X-axis on the graph paper, and the other set is plotted along the vertical or Y-axis. The located points are then connected to form a continuous graph. The action of the oscilloscope in tracing a response curve is so similar to this that some oscilloscopes even have inputs marked "X-amplifier" and "Y-amplifier."

We believe this comparison is a good point to remember. When confusing indications are seen on the oscilloscope screen, it may help if the operator remembers that the oscilloscope is plotting time horizontally and voltage vertically to produce a graphical account of the operating conditions of the circuit. Usually it is unnecessary to know exactly how much time is represented by the horizontal travel of the trace, as long as the beam is uniform in its rate of travel; but, if necessary, this time can be determined accurately and for very short intervals.

Fig. 1-9 shows a graph of one cycle of voltage having a frequency of 60 hertz*. Instantaneous voltage is plotted above or below the X or horizontal axis; elapsed time is plotted to the right of the vertical axis (Y) and measured in fractions of a second. The peak voltage is taken as

---

*One hertz equals 1 cycle per second. Thus, 10,000 hertz (10 kilohertz) equals 10,000 cycles per second.

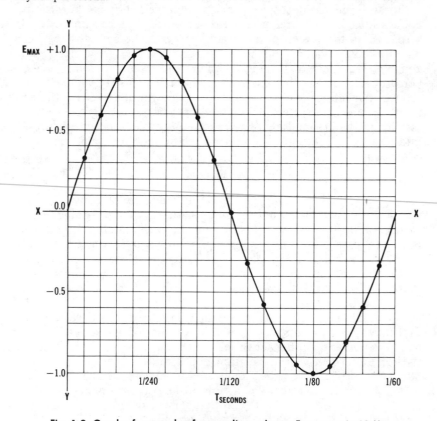

Fig. 1-9. Graph of one cycle of power-line voltage. Frequency is 60 Hz.

1 to simplify plotting the graph. This curve is called a sine curve because the amplitude or Y value at any point on the curve equals the maximum value of E (in this case, 1) times the sine of the X value at that point. (X must be converted to degrees, with one complete cycle equaling 360 degrees.)

Practically all oscilloscopes have a *graticule* mounted in front of the face of the cathode-ray tube. The graticule is commonly ruled like a sheet of graph paper. The vertical intervals are used to measure voltage, and the horizontal intervals are used to measure time in typical applications. Since the plastic graticule is mounted in front of the fluorescent screen in most oscilloscopes, you must view the pattern in a line exactly perpendicular to the screen to minimize parallax error. If you view the screen obliquely, the parallax error can be appreciable in using the graticule. To avoid this source of evaluation error, some modern crt's have parallax-free graticules. For example, the oscilloscope illustrated in Fig. 1-10 has the graticule ruled on the *inner* surface of the crt face.

Even when a parallax-free graticule is utilized in a crt, there remains a practical limit to the precision with which the operator can read the scale. Therefore, some lab-type oscilloscopes are provided with a

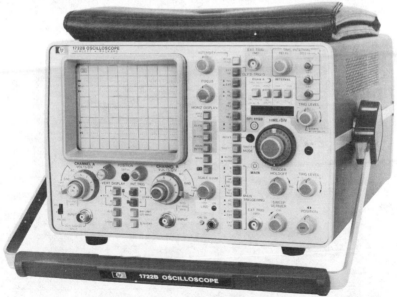

Fig. 1-10. Parallax error is avoided in this oscilloscope by having the graticule ruled on the inside surface of the crt.

digital readout. This function employs *microprocessor (computer) circuitry* to "spell-out" highly precise waveform data automatically, either on an LED panel located below the crt, or as numerals "written" on the crt screen. For example, a crt readout can tell the operator precisely what vertical sensitivity is being used, what time-base speed is in use, what pulse width is being displayed, what the precise rise time may be, what repetition rate is being displayed, and so on. In other words, a crt readout "does the figuring" for the scope operator, and displays a more precise answer. It also reduces measurement setup time, and avoids the possibility of a calculation error. In other words, a digital readout, as illustrated in Fig. 1-11, combines the function of a conventional oscilloscope with that of a digital counter for an elapsed-time indication. In this example, the readout is 1.920 −6, indicating 1.920 microseconds (1.920 × 10^{-6} second).

An oscilloscope with a crt readout features on-screen numerical displays that are produced by scanning action under the control of integrated-circuit character generators. To display a number on the crt screen, the electron beam must be turned on and off during appropriate intervals. It is standard practice to use seven-segment digits, as shown in Fig. 1-12.

The total number of time intervals that are employed for digit formation determines the complexity of the IC character generator(s).

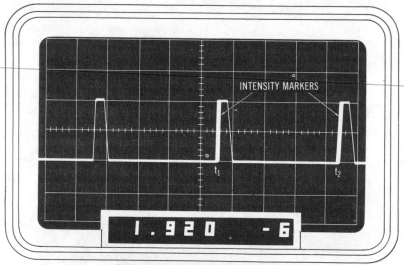

(ELAPSED TIME FROM t_1 TO t_2 IS INDICATED)

Fig. 1-11. An example of a digital readout for the elapsed time between two intensity markers in an oscilloscope pattern.

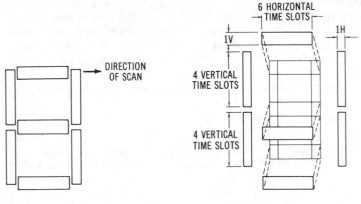

(A) The basic seven-segment digit.

(B) Digit construction.

(C) Time-slot relations.

Fig. 1-12. Operation of the digital readout on an oscilloscope screen.

The seven-segment digits are produced as shown in Fig. 1-12B. Each segment of a digit is comprised of straight lines that fall along either the horizontal or the vertical axis. Thus, four vertical segments occupy two time periods during the horizontal-scanning sequence; they occur at the same time on every line. Note that the three horizontal segments also occupy the same time periods on the horizontal scan. Observe in Fig. 1-12B that the adjoining segments may be proportioned so that their ends overlap, thereby avoiding "breaks" in the character display. Both the horizontal and the vertical scanning periods of each digit are divided into eight *time slots* (Fig. 1-12C). Each of the vertical time slots comprises an even number of horizontal scan lines, in a typical design. The first two horizontal time slots and the first vertical time slot in each digit are blank in this example. The top, center, and lower horizontal

21

segments occupy the third through the eighth horizontal time slots during the second, fifth, and eighth vertical time slots. The left and right vertical segments occupy the third and eighth horizontal time slots. Also, the left and right vertical segments occupy the second through the fifth time slots, and the fifth through the eighth time slots, with an overlap during the fifth time slot.

The more elaborate oscilloscopes often have *edge-lighted* graticules. An example of this design is illustrated in Fig. 1-13. The graticule is fabricated from lucite, and several red-light bulbs are mounted around the edge of the graticule. A control is provided for varying the light intensity. When the scale-illumination control is turned down, the graticule is almost invisible, and only the green fluorescent pattern is in evidence. On the other hand, when the scale-illumination control is turned up, the graticule rulings are displayed as red lines over the green waveform. The adjustable light intensity facilitates the photography of waveforms. The distinction between lab-type and service-type oscilloscopes has become less marked than in the past. State-of-the art service-type scopes have display and operating features that were formerly provided only by lab-type scopes. Also, modern lab-type scopes are far more sophisticated and versatile than their predecessors.

For example, a modern dual-trace triggered-sweep 20-MHz

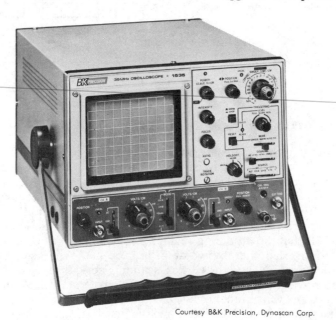

Courtesy B&K Precision, Dynascan Corp.

Fig. 1-13. An oscilloscope with a variable-intensity edge-lighted graticule.

oscilloscope is illustrated in Fig. 1-14. Although this would have been regarded as a lab-type oscilloscope in former years, it now falls in the category of a service-type instrument. A differential input permits the display of the difference between two input waveforms (such as the input and the output of an amplifier), thereby facilitating distortion analysis. The sum of two input waveforms can also be displayed, and an evaluation of propagation delay in digital systems can be progressively effected.

WRITING SPEED

Before discussing the oscilloscope section by section, an important characteristic of all oscilloscopes should be mentioned—the reaction speed of the electron beam to any applied voltage. The beam possesses very little inertia. For all practical purposes, it can be said to have no inertia; consequently, it responds almost instantaneously to the impulse of the deflection voltages. This is the property that enables the trace to follow every variation of the applied signal, no matter how suddenly the signal may change direction or amplitude.

How readily the beam changes direction while moving at high speed can be shown by the following example. Assume an oscilloscope has a sweep frequency of 30 kHz and a horizontal amplification capable of

Courtesy B&K Precision, Dynascan Corp.

Fig. 1-14. A modern dual-trace triggered-sweep oscilloscope.

expanding the trace to four times the screen width. (Many oscilloscopes will exceed both specifications.) For a 5-inch oscilloscope, this means the trace is equal to 20 inches in length although only 5 inches of the center can be seen. The beam sweeps these 20 inches in 1/30,000 second—actually, in even less time since some time is lost in retrace. Thus, the beam is sweeping the tube at a "writing speed" of 600,000 inches per second, or a little faster than 34,000 miles per hour. The retrace time usually is less than trace time; accordingly, the retrace speed would be much greater. However, the retrace is seldom used for viewing and, therefore, is not considered when discussing writing speed.

INPUT IMPEDANCE

Another important characteristic of the oscilloscope is its high input impedance. This is desirable in any voltage-measuring instrument, for it means the instrument will have a minimum loading or disturbing effect on any circuit to which it is connected. The vertical amplifier input impedance of a conventional oscilloscope may have any value from 1 to 5 megohms shunted by 25 to 50 pF. If connected directly to the deflection plates, the impedance may be as high as 10 megohms shunted by 15 pF. The input impedance at the vertical amplifier can be increased by the use of high-impedance probes.

A block diagram of a general-purpose oscilloscope is shown in Fig. 1-15. This is a greatly simplified diagram with several features combined in each section. The focus, intensity, and positioning circuits

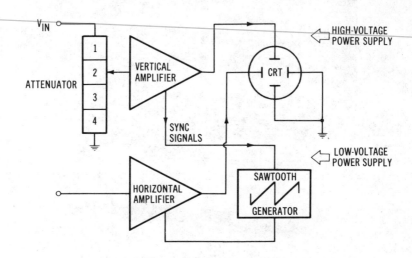

Fig. 1-15. Block diagram of a general-purpose oscilloscope.

are not shown; they have been considered as part of the low-voltage power supply. The step and vernier attenuators usually are associated with the vertical and horizontal amplifiers. Triggering and synchronizing of the sweep oscillator are considered part of the sweep oscillator.

As noted previously, an oscilloscope could be made of a cathode-ray tube and a power supply only. Such an oscilloscope would be extremely limited in the ways it could be used. The signal input would have to be made directly to the deflection plates, and a comparatively strong signal would be necessary to deflect the electron beam a usable amount. After adding vertical and horizontal amplifiers and a horizontal-deflection system to provide a time base, the oscilloscope may be used for an increased number of applications. The oscilloscope can respond to very weak input signals, and general-purpose oscilloscopes sometimes have a vertical-deflection sensitivity of 15 millivolts rms per inch or less.

The appearance of the front panel of a basic oscilloscope is shown in Fig. 1-16. Although numerous controls are provided on the front panel, the instrument is not difficult to operate. To anticipate any subsequent discussion, a half dozen, or even more, of the controls may remain at reference settings during a series of tests. Therefore, the apprentice technician should not jump to the conclusion that an oscilloscope is more difficult to operate than a tv receiver, for example. In fact, a service-type oscilloscope is easier to operate than a color-tv receiver.

When the oscilloscope is properly connected and adjusted, it gives the technician a visible indication of the amplitude, frequency, phase, and waveform of the signal at any particular point in a circuit. An instrument providing as much information as this to a knowledgeable user is a powerful tool indeed. There is probably no phase of electronics where it has not proved useful for designing, testing, or servicing.

CATHODE-RAY TUBE OPERATION

Although a cathode-ray tube operates at a comparatively high voltage (in the range from 1 kV to 5 kV, depending upon the particular design), the crt *draws comparatively little current* from the high-voltage supply, even with the beam-intensity control advanced substantially. Thus, the current demand from the high-voltage power supply in a service-type oscilloscope never exceeds 2 or 3 mA. The normal operating voltages for a crt in a top-performance lab-type oscilloscope are noted in Fig. 1-17.

A modern 5-inch cathode-ray tube is shown in Fig. 1-18. Externally, it has four parts: the base, the neck, the bulb, and the face or screen. Inside the neck, a portion of the gun structure can be seen. Fig. 1-19

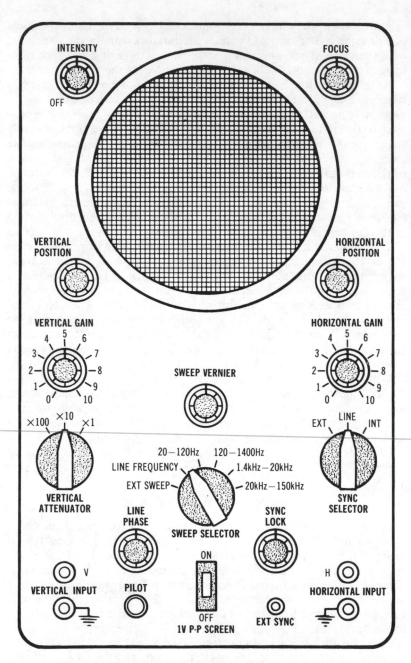

Fig. 1-16. Appearance of the front panel on a basic oscilloscope.

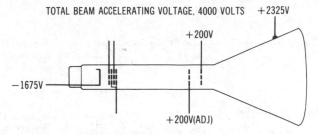

TOTAL BEAM ACCELERATING VOLTAGE, 4000 VOLTS   +2325V

+200V

−1675V

+200V(ADJ)

Fig. 1-17. Terminal voltages for a crt in a high-performance lab-type oscilloscope.

Fig. 1-18. A modern 5-inch cathode-ray tube.

shows the gun structure removed from the tube. The gun contains all the electrodes for forming, shaping, and directing the electron beam that strikes the fluorescent screen of the tube.

Applying the proper voltage to the various electrodes of the gun produces a beam that is brought to a focus in a small spot on the tube screen. The beam intensity is controlled by the voltage on the control grid. The theory pertaining to the focusing action of the gun is probably less interesting to the service technician than the theory pertaining to the action of the deflection plates; consequently, more space will be devoted to the latter. This book will not deal with the electromagnetic deflection systems since they are found almost exclusively in television receivers rather than in oscilloscopes. Fig. 1-20 is a perspective drawing showing how the electron beam passes

Fig. 1-19. The electron gun of a cathode-ray tube.

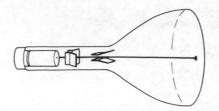

Fig. 1-20. Path of electron beam through deflection-plate assembly.

through the space between the deflection plates on its path to the screen. With all deflection plates at the same electrical potential, the beam will pass along the axis of the deflection-plate assembly and strike the center of the screen.

If one plate of a pair of deflection plates is made more positive or negative than the other, the electron beam is attracted toward the positive plate and repelled from the negative plate (Fig. 1-21), because unlike electrical charges attract and like charges repel each other. The electron beam is always negative and, therefore, is always attracted to the positive plate. The amount of deflection varies directly with the magnitude of the voltage on the deflection plates. For example, if a potential difference of 50 volts between a pair of plates moves the beam 1 inch at the screen, 100 volts will move it 2 inches (Fig 1-22).

Applying an alternating voltage to the vertical plates moves the beam and produces a vertical line from top to bottom of the screen. Similarly, the proper voltage applied to the horizontal plates produces a horizontal line across the screen. With proper voltages for both sets of plates, the beam can be made to move anywhere on the screen.

DEFLECTION SENSITIVITY

The deflection sensitivity of a cathode-ray tube, and of the entire oscilloscope, determines the weakest signal that can be viewed successfully with the instrument. Anyone who has consulted a tube manual about cathode-ray tubes may have noticed that deflection sensitivities can cover a wide range, depending on the voltages used. The sensitivities also differ for the two pairs of deflection plates, one sensitivity being greater than the other. For example, one tube manual lists the following sensitivities for a 5CP1-A cathode-ray tube. When the voltage of anode No. 3 is twice that of anode No. 2, the sensitivity is

Fig. 1-21. The electron beam, being negative, is always attracted by the positive-charged deflection plate and repelled by the negative-charged deflection plate.

28

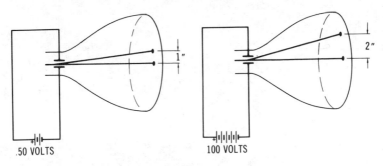

Fig. 1-22. The amount of deflection is directly proportional to the voltage applied to the plates.

39 to 53 volts (dc) per inch for every thousand volts supplied to anode No. 2. This range applies to one set of deflection plates. For the other set under the same voltage conditions, the sensitivity is 33 to 45 volts (dc) per inch per thousand volts supplied to anode No. 2.

The pair of deflection plates having the greater sensitivity (that is, requiring the smaller number of volts per inch of deflection) is always the pair nearest the base of the tube. The reason can be readily seen by examining Fig. 1-23. In this illustration, a pair of deflection plates is shown in two different positions, one (position A) being nearer the tube base than the other. If the applied voltage is the same for both positions, the electron beam will be deflected through an equal angle each time. With equal deflection angles, the deflection plates at position A swing a longer beam, thus giving a longer trace.

Fig. 1-23. Given equal deflection voltages, deflection plates set nearest the base of the cathode-ray tube will have a greater deflection sensitivity than those farther away.

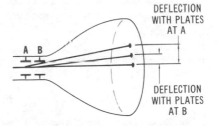

Either pair of plates can be used for the vertical system; the rotational position of the tube about its long axis determines which pair. To obtain the highest possible deflection sensitivity for the vertical system, the cathode-ray tube normally is so positioned that the pair of plates closest to the base produce the vertical deflection. The horizontal-deflection plates usually are driven by a stronger signal and are farther from the base than the vertical-deflection plates.

29

CHAPTER 2

Power Supplies

GENERAL REQUIREMENTS

Power supplies for oscilloscopes have a high-voltage section and a low-voltage section. In some design, a negative-output high-voltage arrangement is used in order to simplify safety requirements. In other designs, both negative-output and positive-output high-voltage configurations are utilized. As noted previously, a crt high-voltage power supply does not provide substantial current. On the other hand, the low-voltage power supply must provide adequate current for the operation of the amplifiers, time base, and any supplementary subsections in the scope. The low-voltage power supply may have positive output, negative output, or both positive and negative outputs. The more elaborate designs of oscilloscopes employ *regulated power supplies* for stability of operation. Nearly all modern oscilloscopes have solid-state low- and high-voltage power supplies.

WARNING

Cathode-ray tubes in oscilloscopes operate at high voltage, which may cause injury or death if accidentally contacted. Only experienced technicians should be permitted to operate an oscilloscope with its case removed. A high-voltage power supply in an oscilloscope can be deceptive, because the crt heater and cathode may be operated with 1000 volts or more above ground. Note that even when the power switch to an oscilloscope is turned off, the filter capacitors may retain a deadly charge for a long time, if the high-voltage bleeder resistor is open-circuited. Therefore, the high-voltage filter capacitors should always be discharged before starting to troubleshoot oscilloscope circuitry. Keep in mind also that defective circuits may cause dangerously high voltages to appear at unexpected points in oscilloscope networks.

ELECTRON PATH THROUGH A CATHODE-RAY TUBE

Fig. 2-1 shows a 5UP1 cathode-ray tube as it commonly appears in the schematic diagrams of oscilloscope instruction books. The order indicated for the elements, starting from the base of the tube, is the same in the diagram as for the actual tube, except possibly anode No. 2. Anode No. 2 is connected internally to grid No. 2. It is also connected to the coated interior of the bulb of the tube, although this is not shown in the diagram. The coating extends almost to the face of the screen and accelerates the electron beam on its way to the screen. It also collects the electrons of the beam after they have struck the screen.

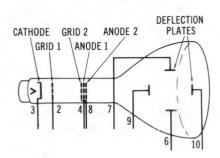

Fig. 2-1. Control elements in a 5UP1 cathode-ray tube.

The path of the electrons through the cathode-ray tube is as follows. The electrons are emitted from the heated cathode and, being negative, are attracted toward the nearest positive element, grid No. 2. They pass through apertures in grid No. 2 and also in anodes No. 1 and No. 2, and are subjected to the action of the deflection plates. Finally, they strike the screen of the tube causing a spot or trace of light, and they are then collected by the interior coating which forms a part of anode No. 2. Thus, the electron path through the tube originates at the cathode and terminates at anode No. 2.

Polarities of the Tube Elements

Proper operation of the cathode-ray tube requires anode No. 1 to be more positive than the cathode and anode No. 2 to be more positive than anode No. 1. Grid No. 1 is the control grid and operates at a voltage equal to or more negative than that of the cathode. Its action is similar to that of the control grid of a receiving tube—it controls the number of electrons flowing between the cathode and anode. Being negative, it repels the negative electrons, and if it becomes negative enough, the electron beam is cut off entirely.

The intensity control of the oscilloscope is usually connected to grid No. 1 of the cathode-ray tube, although it may be connected to the cathode instead. A variable intensity depends on a variable potential difference between the cathode and the grid. This variable potential difference can be obtained if the potential of one element is varied while the potential of the other is held constant. The technician is familiar with this aspect of the operation of the cathode-ray tube through his association with television receivers. In some receivers, the picture-tube element to which the brightness control is connected is the control grid; in others, it is the cathode.

Range of Voltage for Normal Operation

As was stated previously, the necessary potentials for operation of the cathode-ray tube are furnished by the high-voltage section of the power supply. These potentials can vary over a wide range, and satisfactory operation will still be obtained. For example, the voltage at anode No. 2 of a 5UP1 tube can be from 1000 to 2500 volts with respect to the cathode. Operation below 1000 volts is not recommended. No matter which voltage is chosen, there are some advantages and disadvantages. The lower voltages can be attained more easily and economically, and they make possible a higher deflection sensitivity. These advantages are offset by less brilliance of the spot and poorer focusing qualities.

BASIC POWER-SUPPLY CONFIGURATIONS

Various rectifier arrangements are used in oscilloscope power supplies, as shown in Fig. 2-2. The chief distinction between half-wave and full-wave operation is that the latter has double the ripple frequency of the former. However, a higher ripple frequency can be filtered with a less elaborate filter network. Oscilloscope power supplies often use bridge rectifier configurations, as depicted in Fig. 2-3. The chief advantage of the basic bridge arrangement is that the full secondary voltage is rectified, whereas, the full-wave arrangement rectifies only half of the total secondary voltage. Note that a full-wave/full-wave combination configuration provides both positive and negative outputs, with a ripple frequency that is double the supply frequency. In high-voltage supplies, it is a common design practice to connect rectifier diodes in series, as exemplified in Fig. 2-3D. Using a series connection increases the peak-inverse voltage that is permissible in the high-voltage rectifier-filter system.

Some lab-type oscilloscopes may employ high-voltage power supplies that include a 60-kHz oscillator, as shown in Fig. 2-4. In this design, a conventional power supply is followed by an oscillator that generates a 60-kHz sine wave which, in turn, energizes the high-voltage power supply. The basic advantage of this type of arrangement is that comparatively small high-voltage filter capacitors

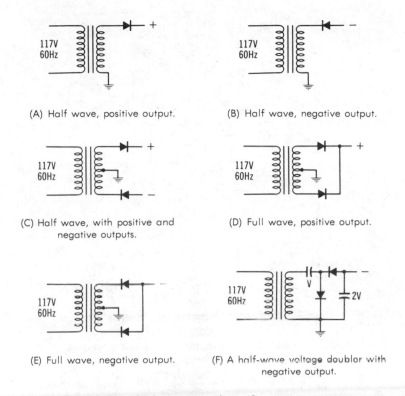

(A) Half wave, positive output.

(B) Half wave, negative output.

(C) Half wave, with positive and negative outputs.

(D) Full wave, positive output.

(E) Full wave, negative output.

(F) A half-wave voltage doubler with negative output.

Fig. 2-2. Some basic power-supply rectifier arrangements.

can be used—large high-voltage filter capacitors are quite costly. Note that most state-of-the-art oscilloscopes utilize an RC filter circuitry; filter chokes are both expensive and heavy components. As explained next, the regulated power supplies used in modern oscilloscopes also provide an incidental filtering function, so that the elaborate LCR filter circuitry used in earlier designs is not required.

POWER-SUPPLY SYSTEMS

A configuration for a power supply that is used in a triggered-sweep oscilloscope is shown in Fig. 2-5. It consists of a low-voltage section and a high-voltage section. The line voltage is connected through a slow-blow fuse and a power switch to the primary windings of the power transformer. The dual primary-windings of the transformer may be either connected in parallel for 120-volt operation, or in series for 240-volt operation. The high-voltage secondary winding of the power transformer is connected to the voltage-doubler circuit

33

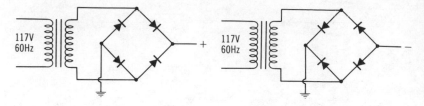

(A) Full-wave bridge, positive output.   (B) Full-wave bridge, negative output.

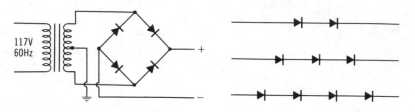

(C) A full-wave/full-wave combinational circuit with positive and negative outputs.

(D) Rectifier diodes may be connected in series in order to withstand a greater peak-inverse voltage.

Fig. 2-3. Some basic bridge rectifier configurations.

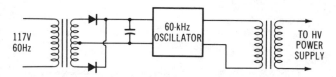

Fig. 2-4. The ripple frequency of the high-voltage power supply is 1000 times higher than that of the 60-Hz source.

consisting of D401, D402, C403, and C404. Resistor R403 and capacitor C402 filter this negative high voltage, which is then coupled through resistor R502 to the grid of the crt. The intensity and focusing voltages are also supplied to the crt, from the voltage-divider network that consists of resistors R404, R505, R504, R503, and R402. A separate 6.3-volt winding supplies the crt heater voltage.

Two secondary transformer windings supply graticule lighting and calibration voltages to the front panel. One of these windings supplies a 1-volt peak-to-peak potential to the 1-V p-p jack on the front panel. The other winding supplies the necessary voltage to illuminate the graticule lights, the intensity of which is varied by the graticule control R507. The low-voltage secondary winding is connected to the full-wave rectifier diodes D405, D406, D407, and D408. Zener diode ZD404 and resistor R408 maintain a constant voltage to the base of

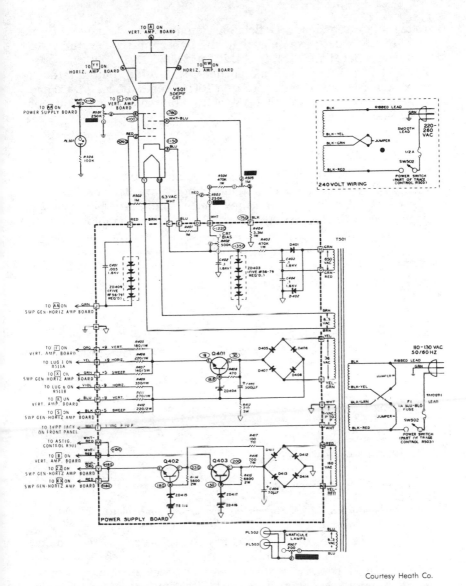

Courtesy Heath Co.

Fig. 2-5. Power supply for a triggered-sweep oscilloscope.

pass transistor Q401. The output from the series pass transistor is a regulated 36 volts. Since a regulator operates to maintain a constant output voltage, it has considerable incidental filtering action, and it also provides a very low internal impedance. By connecting equal loads from each side of the supply to ground, six separate dc-output voltages are obtained. Deflection potentials are obtained from another secondary winding that is connected to the full-wave bridge-rectifier diodes D411, D412, D413; and D414. Zener diodes ZD417 and ZD418 provide a regulated +150-volt dc output through transistor Q403 and dropping resistor R416. Zener diodes ZD415 and ZD416 provide a regulated +180-volt dc output through transistor Q402 and dropping resistor R417.

VOLTAGE REGULATION

The modern trend is to use regulated power supplies in oscilloscopes. This makes for stability of operation and more accurate measurements of voltage and time under conditions of line-voltage fluctuation. The simplest voltage-regulating circuits employ voltage-regulator tubes, as depicted in Fig. 2-6. When the line voltage fluctuates, the current flow through the voltage-regulator tube changes, but the voltage drop across the tube remains practically constant.

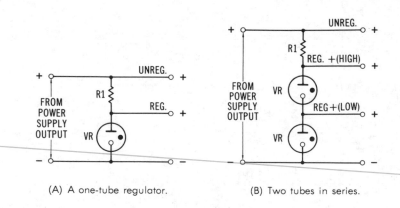

(A) A one-tube regulator.   (B) Two tubes in series.

Fig. 2-6. Basic configuration for voltage-regulator tubes.

Fig. 2-7 shows the basic configuration for a zener diode voltage-regulator circuit. From a practical viewpoint, the action of a zener diode is the same as that of a voltage-regulator tube. In other words, when the supply voltage fluctuates, the current flow through the zener diode changes, but the voltage drop across the diode remains practically constant. This action results from the fact that a zener diode does not conduct current in its reverse-biased direction until the applied voltage rises to a critical value. At this critical value, "breakdown" occurs and the internal resistance of the zener diode becomes very small. Note resistor R in Fig. 2-7; this is a

36

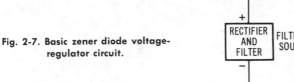

Fig. 2-7. Basic zener diode voltage-regulator circuit.

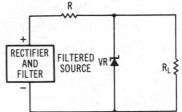

current-limiting resistor which prevents the zener diode from drawing so much current from the source that the diode would be damaged. Note also that the load resistance, RL, may vary over an appreciable range, and the voltage across the zener diode will remain practically unchanged.

You will find transistorized regulated power supplies in many lab-type scopes. The principle of operation is basically the same as in tube-type regulated power supplies. In other words, the voltage reference is provided by a voltage-regulator tube or a zener diode. If the current demand on the power supply increases, the transistors "sense" that the output voltage is decreasing with respect to the reference voltage. In turn, the base-emitter bias automatically changes to permit more current flow and thereby maintain the output voltage at a practically constant value.

In a regulator circuit, the difference between a reference input (e. g., the supply voltage) and some portion of the output voltage (e.g., a feedback signal) is used to supply an error signal to the control elements. The amplified error signal is applied in a manner that tends to reduce this difference to zero. Regulators are designed to provide a constant output voltage that is very nearly equal to the desired value in the presence of a varying input voltage and an output load.

In series regulator circuits, such as that shown in Fig. 2-8, direct-coupled amplifiers are used to amplify an error or difference signal that is obtained from a comparison between a portion of the output voltage and a reference source. The reference-voltage source, V_R, is placed in the emitter circuit of the amplifier transistor Q1 so that the error or difference signal between V_R and some portion of the output voltage, V_o, is developed and amplified. The amplified error signal forms the input to the regulating element consisting of transistors Q2 and Q3.

In many situations, a device for a high-voltage power supply is available with sufficient voltage capability but with insufficient current dissipation or second-breakdown capability. The series-regulator circuit shown in Fig. 2-9 solves this problem by reducing the dissipation and current requirements in the high-voltage device Q1.

Shunt regulator circuits are not as efficient as series regulator

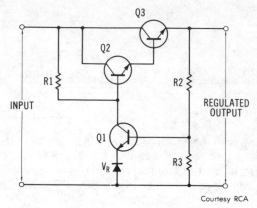

Fig. 2-8. A basic series regulator circuit.

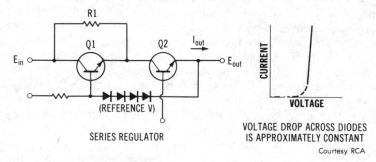

Fig. 2-9. High-voltage power-supply regulation.

circuits for most applications, but they have the advantage of greater simplicity. In the shunt voltage-regulator circuit shown in Fig. 2-10, the current through the shunt element, consisting of transistors Q1 and Q2, varies with changes in the load current or the input voltage. This current variation is reflected across resistance R1, in series with the load, so that the output voltage, V_o, is maintained nearly constant. In

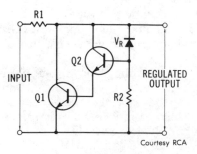

Fig. 2-10. A basic shunt regulator circuit.

Fig. 2-10, diode V_R provides a constant voltage reference for development of the error voltage.

NEGATIVE HV-SUPPLY FEATURES

Most oscilloscopes use the arrangement shown in Fig. 2-11B for the high-voltage supply, often with minor variations. For example, in Fig. 2-12, the rectifiers and filter sections have been omitted; the final stages and the positioning controls for one of the amplifier channels are shown. R1 and R2 form the ground return for one pair of deflection plates, and the ac output from Q1 and Q2 is developed across these two resistors. This is a push-pull deflection system in which a negative-going signal is applied to one deflection plate of a pair at the same time that a positive-going signal is applied to the other plate.

With the circuit arrangement shown in Fig. 2-12, the dc potential of either deflection plate will not vary greatly from ground potential. Any variation will be due to the action of positioning controls R3A and R3B. These controls are ganged and so wired that any rotation of the common shaft shifts the slider of one control toward a more positive potential and, at the same time, shifts the slider of the other control toward a more negative potential. As a result, the deflection plates have a push-pull action for the dc positioning voltage as well as for the ac signal.

The following advantages result from a negative high-voltage supply:

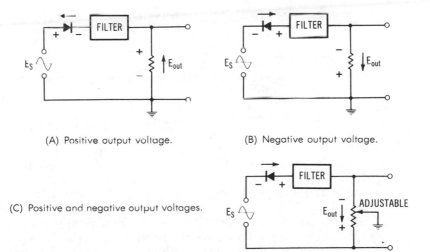

(A) Positive output voltage.

(B) Negative output voltage.

(C) Positive and negative output voltages.

Fig. 2-11. Several variations of a half-wave rectifier circuit.

39

1. The deflection plates can be operated at a dc potential close to that of anode No. 2, thus eliminating the defocusing effect obtained when the two potentials differ greatly.
2. Capacitors C1 and C2 can have a fairly low voltage rating.
3. The circuit can be more easily adapted to dc connection between deflection plates and amplifiers.
4. Less insulation is needed between the positioning controls and the chassis or the front panel.

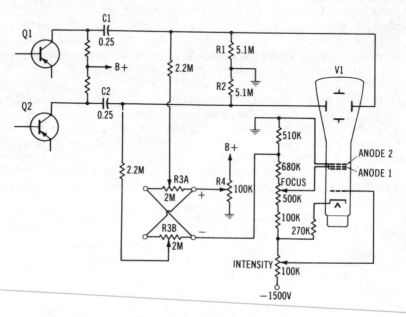

Fig. 2-12. Partial schematic showing the positioning controls and the high-voltage divider network of an oscilloscope.

Contrast the preceding conditions with those obtained if the polarity of the high-voltage supply were reversed.

1. Anode No. 2 would be at a high positive potential to ground, resulting in an extreme difference in potential between the deflection plates and anode No. 2 if dc connections are made from the amplifier to the deflection plates. (This condition is undesirable.)
2. If blocking capacitors C1 and C2 are used, the deflection plates and anode No. 2 would be at nearly the same potential, but the voltage rating of the capacitors would have to be high. Capacitors of that value and rating would be bulky and expensive.
3. The horizontal- and vertical-positioning controls would have to

be highly insulated from the chassis and the front panel to protect against the high voltage.

Regardless of the polarity of the high-voltage supply, the voltage rating of filter capacitors C403 and C404 in Fig. 2-5 must be high. In summary, the advantages seem to lie mainly with a high-voltage supply of negative polarity, and most oscilloscopes employ this system.

INSULATION OF FRONT PANEL CONTROLS

As can be seen in Fig. 2-12, the focus and intensity controls are at points of fairly high potential at the negative end of the dividing network. Consequently, the manufacturers take precautions to insulate these controls from the chassis and the front panel. The method used in one oscilloscope is shown in Fig. 2-13. Insulating washers are used between each control and the mounting bracket, with an insulating coupler between the control shaft and the long metal shaft running to the front panel.

Fig. 2-13. One method of insulating between a control and the chassis.

BEAM INTENSIFICATION

Because it is popular with oscilloscope manufacturers, the 5UP1 cathode-ray tube has been used as an example. However, other cathode-ray tubes are also found in oscilloscopes, and some require a power supply slightly different from those discussed so far. An intensifier anode (called anode No. 3) in the 5ABP1 tube and 5CP1A tube may sometimes be operated at a potential as much as 2000 volts positive with respect to ground; and at the same time, the control grid may be as much as 2000 volts negative with respect to ground. The intensifier anode greatly accelerates the electrons in the beam after they have passed between the deflection plates. A brighter spot results, yet the deflection sensitivity is not seriously affected. The increased velocity of the electrons in the beam permits a higher scanning rate. In order to obtain the positive high voltage for the intensifier anode, another half-wave rectifier system can be added.

With all of these high-voltage sources present within the case, the operator should be extremely careful when examining the interior of an oscilloscope. The instruction manuals caution against operating the oscilloscope with the chassis outside its case. Before touching any part of the interior of an oscilloscope, the operator should make sure that the filter capacitors are not charged.

CHAPTER 3

Sweep Systems

OVERVIEW OF OSCILLOSCOPE OPERATING FEATURES

In Chapter 1, it was mentioned that the oscilloscope will actually plot a graph of voltage ("write" a pattern) with respect to time on the crt screen. The oscilloscope operator can see the way in which this voltage changes in amplitude and direction from one moment to the next. The signal to be observed and/or measured is normally applied to the vertical-deflection channel and will cause a vertical trace to be displayed on the crt screen, provided that the signal is of sufficient amplitude and that no ac voltage is applied to the horizontal-deflection plates (see Fig. 3-1).

Under these conditions, a change of amplitude in the signal will result in a change of the height of the vertical trace. In order that these changes in amplitude may be viewed with respect to changes in time, some type of sweep system must be incorporated in the oscilloscope.

Fig. 3-1. Application of an ac signal to the vertical channel of an oscilloscope produces a vertical-line display on the crt screen.

The signal from the sweep system is used to drive the horizontal-deflection plates of the cathode-ray tube. This provides a horizontal trace as a time reference for the signal at the vertical-deflection plates. Because of this, sweep systems are sometimes called time bases. In addition to the sweep signals provided internally in the general-purpose oscilloscope, other sweep signals can usually be applied from an external source.

Oscilloscope sweeps may be classed as linear or nonlinear, and as single or repetitive. Single sweeps have a great utility for viewing signals of a nonrecurring nature. They are designed to sweep the beam once across the screen of the oscilloscope and they must be timed accurately so that the signal to be viewed will occur at the exact instant of the sweep. A sweep of such short duration would result in a trace that would fade very quickly on a screen of normal persistence; consequently, a screen with a long persistence is used to increase the viewing time.

The majority of the signals that a service technician will encounter are of a recurring nature. They normally go through a complete cycle of variations, a number of times a second. Some examples of this type of signal are:

1. The voltage supplied by the power line.
2. The ac voltages at tube filaments in a receiver.
3. The voltages generated by the sweep circuits in a television receiver.

The ideal sweep for viewing these signals is one in which the beam starts at the left-hand edge of the oscilloscope screen and moves at a uniform rate of speed in a horizontal direction to the right edge of the screen. When it reaches the right edge, the sweep should reverse direction and return to the starting position at the left of the screen. This return sweep (called retrace) should be made in the least time possible.

OSCILLOSCOPE CONTROLS

Various oscilloscope controls have been previously discussed. In a modern triggered-sweep oscilloscope, specialized time-base (sweep) controls are provided, as exemplified in Fig. 3-2. The function and operation of these controls and indicators will become clear as the reader proceeds with the following explanations. Follow the numbers given in the illustrations of Figs. 3-2 and 3-3.

    1 GRATICULE scale. This scale provides calibration marks for voltage measurements. Only the left (0 to 1) vertical scale or the right (0 to 3) vertical scale is illuminated with respect to the voltage range being used for CALI-BRAIN.

    2 VOLTS FULL SCALE indicator (.100, 1.00, 10.0, and 100). This

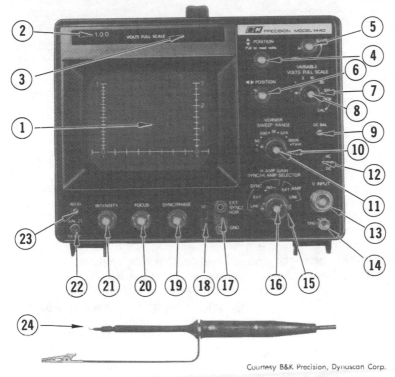

Courtesy B&K Precision, Dynascan Corp.

Fig. 3-2. An example of triggered-sweep oscilloscope controls and functions.

indicator indicates full scale peak-to-peak voltage reading of the graticule scale that is directly below the indicator. The indicator is illuminated only when switch **7** is set to the .1, 1, 10, or 100 volt range. The decimal point moves automatically with the VOLTS FULL SCALE selector switch **7**. The vertical VARIABLE control **8** must be set to CAL for the indicated value to be correct.

3 VOLTS FULL SCALE indicator (.300, 3.00, 30.0, and 300). Same as indicator **2**, except it is automatically illuminated when switch **7** is set to the .3, 3, 30, and 300 volt ranges.

4 ↕POSITION/PULL TO READ VOLTS switch. Rotation of this switch adjusts vertical position of trace. This push-pull switch activates CALI-BRAIN when pulled out, and is in normal operation when pushed in. With CALI-BRAIN activated, horizontal deflection ceases and the resulting vertical deflection produces a thin vertical line that is displayed adjacent to the graticule scale on which the value is to be read. On the .1, 1, 10, and 100 volt ranges, the line appears adjacent to the left vertical graticule

45

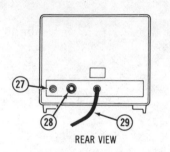
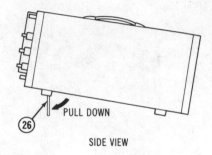

REAR VIEW   SIDE VIEW

Fig. 3-3. Sketch showing other operator's facilities for the oscilloscope
illustrated in Fig. 3-2.

scale, and on the .3, 3, 30, and 300 volt ranges, the line appears adjacent to the right vertical graticule scale.

5 OFF/ILLUM control. In the full counterclockwise click-stop position, ac power to the oscilloscope is OFF. Initial clockwise rotation turns the unit ON, and any further rotation reduces the illumination of the graticule scale and the VOLTS FULL SCALE indicators 2 and 3.

6 ◄►POSITION control. This control adjusts horizontal position of the trace; becomes disabled when CALI-BRAIN is operative.

7 VOLTS FULL SCALE selector switch (vertical attenuator). This switch provides coarse adjustment of the vertical sensitivity. Vertical sensitivity is calibrated in 8 steps, from .1 volt to 300 volts full scale, when VARIABLE 8 is set to the CAL position. The switch also automatically selects the graticule scale and the scale indicator, including the correct placement of the decimal point for the CALI-BRAIN display.

8 VARIABLE (vertical attenuator). This allows vernier adjustment of the vertical sensitivity. In extreme clockwise (CAL) position, the vertical attenuator is calibrated. The control must be in the CAL position for CALI-BRAIN display measurements.

9 DC BAL adjustment. This is the vertical dc balance adjustment.

10 SWEEP RANGE selector switch. This switch selects the coarse horizontal sweep rates in 5 steps—from 5 Hz to 500 kHz plus TV-V and TV-H positions. The selector positions fall between two numbers which indicate the range of sweep rates that may be selected with the VERNIER control 11. For example, the first position provides sweep rates of 5 Hz to 50 Hz, depending upon the setting of the VERNIER control. The TV-V (television vertical) and TV-H (television horizontal) positions provide the correct sweep rates for viewing two full fields and two full lines of television composite video waveforms.

11 VERNIER control. This control provides the fine horizontal sweep rate adjustment. In the fully counterclockwise position, the sweep rate equals the lowest rate of the selected SWEEP RANGE **10**. In the fully clockwise position, the sweep rate equals the highest rate of the selected SWEEP RANGE.

12 AC-GND-DC switch. This is the vertical input selector switch. In the AC position, it blocks the dc component of the input signal. In the GND position, it opens the signal input path and grounds the input to the vertical amplifier. This provides a zero-signal base line, the position of which can be used as a reference when performing dc measurements. In the DC position, the direct input of the ac and dc component is allowed.

13 V INPUT jack. This vertical input is a uhf-type connector; it also accepts a banana-type plug in the center terminal.

14 GND terminal. This is chassis ground; it accepts wire, pin, or banana-type plug.

15 SYNC/H. AMP SELECTOR switch.

   1. SYNC positions are:

     LINE—The power-line frequency sync (50/60 Hz).

     EXT—The signal at EXT SYNC/IIOR jack **17** is used for sweep synchronization.

     + INT—The positive portion of the waveform being observed is used for sweep synchronization.

     − INT—The negative portion of the waveform being observed is used for sweep synchronization.

   2. AMP positions are:

     EXT—The sweep generator is disabled and the external signal applied at EXT SYNC/HOR jack **17** provides horizontal deflection.

     LINE—The line-frequency (50/60 Hz) sinusoidal wave is used for horizontal deflection; SYNC/PHASE control **19** adjusts the phase of the sweep.

16 H AMP GAIN control. This adjusts the horizontal size of the display.

17 EXT SYNC/HOR jack. This is the external sync or external horizontal input.

18 DC-AC switch. The horizontal input selector switch. In the DC position, there is a direct input of the ac and dc component. In the AC position, it blocks the dc component of the horizontal input signal.

19 SYNC/PHASE control. This control adjusts the sync signal level for proper waveform synchronization; it doubles as a phase control when using line sweep.

20 FOCUS control. This control adjusts sharpness of trace.

21 INTENSITY control. This control adjusts brightness of trace.

22 Cal 〰️ jack. This provides a calibrated 1-volt peak-to-peak square, with fast rise-time at line frequency, for checking scope calibration and compensating the 10:1 probe mode.

23 Astig adjustment. Used in conjunction with focus 20, this control adjusts the overall sharpness of the trace.

24 Probe. This is a combination 10:1 attenuation and direct measurement probe for the vertical input.

25 Vector overlay. A plastic overlay that may be inserted over the screen of the cathode-ray tube for vectorscope measurements (not illustrated).

26 Tilt stand. With stand folded up, the scope sits on rubber feet. Unfolded, the front of the scope is elevated to a convenient viewing height.

27 Int mod jack. An intensity modulation (Z axis) input.

28 Fuse. A 1-amp, 3-AG–type fuse.

29 Power cord. A 3-conductor grounding-type cord; it is ground-connected directly to the chassis for maximum safety.

LINEAR SAWTOOTH SWEEP

The waveform of the voltage that is necessary to produce a linear sawtooth sweep is shown in Fig. 3-4. Several cycles of the sawtooth waveform are shown. The voltage applied to the horizontal deflection plates is plotted in a vertical direction, and time is plotted in a horizontal direction. The sweep produced by such a waveform is called a linear sweep because the useful portion of it moves at a constant rate of speed and can be represented by a straight line on a graph. In many oscilloscopes, the retrace is blanked out and does not appear on the screen.

Retrace blanking can help prevent some of the confusing indications that might be seen without blanking. This is especially true where some of the higher sweep frequencies are used. Usually, raising the sweep rate higher and higher will result in sweep voltage cycles that contain a

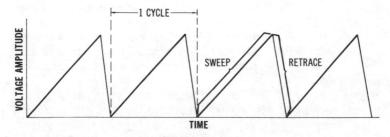

Fig. 3-4. Voltage waveform used to display a pattern on an oscillosope screen.

48

larger percentage of retrace time. If the retrace is permitted to appear on the screen together with any vertical deflection caused by a signal at that time, it may obscure some more important detail that is occurring during the forward portion of the sweep.

Blanking can be accomplished by applying the retrace signal from the sweep generator circuits to either the cathode or grid of the cathode-ray tube for intensity modulation. Circuits may be inserted between the generator and the cathode-ray tube to provide any waveshaping, phase shifting, or amplification that may be necessary.

The waveform of Fig. 3-4 can be approximated very closely by the voltage across a capacitor being charged and discharged in a certain manner. Fig. 3-5 shows a simple arrangement for doing this. When the switch is in position A, capacitor C will be shorted, and no voltage will appear across its terminals. When the switch is moved from point A to point B, the battery will immediately start to charge the capacitor and will continue to charge capacitor C until the voltage across the capacitor equals that across the battery. Theoretically, it would take an infinite length of time for E_C to reach the voltage E_B. For most practical purposes, E_C can be considered to equal E_B after a time equal to 5RC has elapsed. RC time is measured in seconds and is equal to the product of the resistance in megohms times the capacitance in microfarads.

Fig. 3-6 is a graph showing the ratio between the voltage E_C and the voltage E_B obtained with the circuit of Fig. 3-7. It can be seen that the voltage E_C increases rapidly at first, then more slowly as E_C approaches E_B. Considered as a whole, the curve of Fig. 3-6 appears to have a large amount of curvature. However, if only a small portion of the curve is

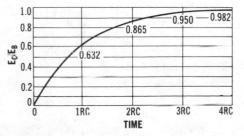

Fig. 3-5. A simple arrangement for charging and discharging a capacitor.

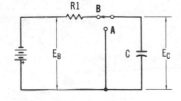

Fig. 3-6. Graph showing the rise in voltage as a capacitor is charged through a resistor.

49

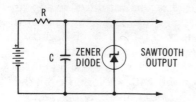

ZENER DIODE

SAWTOOTH OUTPUT

Fig. 3-7. An elementary sawtooth-generator circuit.

considered at one time, it appears to be nearly straight, especially between points 0 and 1RC. It would, therefore, be logical to use this latter portion of the curve, or a part of it, to develop the sawtooth curve diagrammed in Fig. 3-4.

With reference to Fig. 3-6, consider how a repetitive sawtooth waveform is generated. The battery charges capacitor C through resistor R, in accordance with the exponential curve shown in Fig. 3-6. When the breakdown voltage of the zener diode is reached, the capacitor suddenly discharges through the diode. Thereupon, the cycle of operation is repeated. If R and C have large values, the repetition rate is slow; on the other hand, if R and C have small values, the repetition rate is high. Note that if the value of R or the value of C is adjustable, the sawtooth repetition rate will be controllable. The RC combination has a *time constant*; its time constant in seconds is equal to ohms times farads. For example, if R is 1 megohm and C is 1 microfarad, the circuit time constant is one second. This means that the sawtooth voltage will rise to 0.632 of its maximum possible value in 1 second. Refer to Fig. 3-8 and you will see that the sawtooth output becomes more linear if the source voltage is increased. In other words,

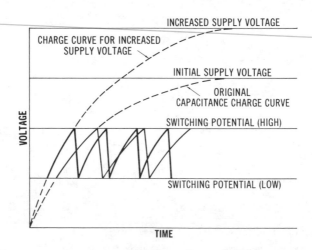

Fig. 3-8. Frequency and linearity changes caused by variation in the supply voltage.

the linearity is improved because the high and low transition points are then farther down on the total charge curve, where the curve is more nearly linear.

SWEEP RATE CONTROL

A wide and continuous range of sawtooth repetition rates must be provided in a practical horizontal-sweep (time-base) system. Thus, operating controls are provided for adjusting the time constant of the integrating circuit that forms the ramp (sawtooth). This is usually accomplished with a variable resistance (potentiometer) control for fine or vernier frequency variation, and with a graduated series of fixed capacitors connected to a switch control for coarse (step) frequency variation. A rotary switch is usually employed for the sweep selector (step-frequency control). At this point, it is helpful to briefly consider basic rectangular-wave generators, such as those that are widely utilized in time-base arrangements. One basic type is free running, and the other is actuated by a trigger pulse.

A basic transistor multivibrator circuit is shown in Fig. 3-9. This free-running (astable) multivibrator is essentially a nonsinusoidal two-stage amplifier with its output coupled back to its input. As a result, one transistor conducts while the other is cut off, until a point is reached at which the stages reverse their conditions. In other words, the stage that had been conducting cuts off, and the stage that had been cut off conducts. This oscillatory action is used, for example, to produce a square-wave output. By suitable modification, the circuit can be made to produce a sawtooth waveform output.

The foregoing discussion of multivibrator circuit action is sum-

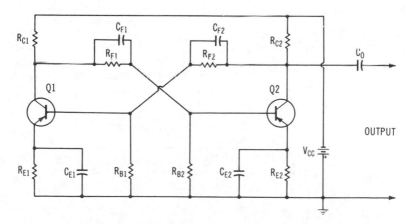

Fig. 3-9. Basic transistor multivibrator circuit.

marized by the waveforms depicted in Fig. 3-10 for the circuit of Fig. 3-9. Note that i_{b1} denotes the base current of Q1, v_{b1} denotes the base voltage of Q1, i_{c1} denotes the collector current of Q1, v_{c1} denotes the collector voltage of Q1, i_{b2} denotes the base current of Q2, v_{b2} denotes the base voltage of Q2, i_{c2} denotes the collector current of Q2, and v_{c2} denotes the collector voltage of Q2. Although the collector waveforms are semisquare and have no resemblance to a sawtooth wave, we will find that comparatively simple circuit modifications can provide a collector-voltage waveform that is a good approximation of a sawtooth waveform.

TRIGGERED SWEEP

Conventional scopes use a free-running sawtooth oscillator for horizontal deflection. *Triggered-sweep* scopes use a sawtooth generator that is not free running. Therefore, a sawtooth waveform is generated only when a vertical-input signal is applied. The leading edge of the input signal (Fig. 3-11) triggers the sawtooth generator, and one sweep excursion occurs. The sawtooth generator then remains inactive until the next leading edge arrives. This means that before the

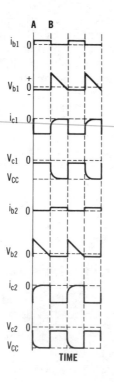

Fig. 3-10. Voltage and current waveforms in a basic transistor multivibrator circuit.

52

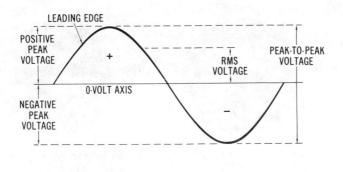

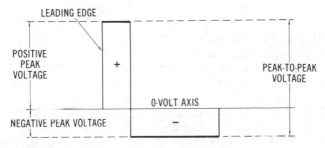

Fig. 3-11. Leading edges in sine and pulse signal waveforms.

leading edge of a vertical-input signal arrives, you see only a spot on the screen, as illustrated in Fig. 3-12A. When the leading edge arrives, the beam deflects horizontally, as seen in Fig. 3-12B. In turn, a waveform can be greatly expanded merely by advancing the horizontal sweep-rate control (Fig. 3-13). Trigger action is obtained by biasing at least one of the tubes or transistors in a multivibrator beyond cutoff. For example, in a cathode-coupled multivibrator, a variable common-cathode bias control may be provided to set the triggering level.

Some oscilloscopes designed for industrial applications have provision for very slow horizontal sweep. A binding post may be provided on the front panel, marked "external capacitor." When a capacitor is connected between the binding post and ground, the time constant of sweep oscillator is increased. In turn, the horizontal-deflection rate slows down. The scope manufacturer often supplies a chart that shows the sweep rate versus the external capacitance value. A triggered sweep, or time base, uses the basic monostable (one-shot) multivibrator circuit shown in Fig. 3-14A. This configuration is similar to the multivibrator arrangement described in the preceding section, except that the bias voltage V_{BB} holds Q1 below cutoff, whereas, the supply voltage V_{CC} holds Q2 in saturation (conduction) during the time

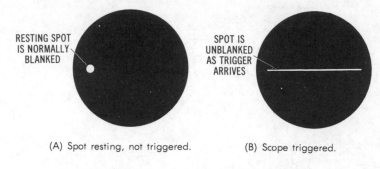

RESTING SPOT
IS NORMALLY
BLANKED

SPOT IS
UNBLANKED
AS TRIGGER
ARRIVES

(A) Spot resting, not triggered.

(B) Scope triggered.

Fig. 3-12. Trigger control of the cathode-ray tube beam.

that no trigger (sync) voltage is applied to Q1 via C_C. The sequence of multivibrator action is seen in Fig. 3-14C. When a negative sync voltage (trigger pulse) is applied to the base of Q1 (see v_{in}), the transistor is momentarily driven into conduction. In turn, the trigger pulse is amplified and applied to the base of Q2 in positive polarity. Thus, Q2 is cut off (see v_{b2}) and Q1, in turn, is held in conduction until the charge on C_{F1} decays through R_{F1}. At this point, Q2 goes back into saturation and Q1 goes back below cutoff. The result is a single square output pulse (see v_{c2}). The multivibrator then rests in its quiescent state until another trigger pulse is applied to the base of Q1.

Before this square output pulse can be used to deflect the crt beam, it must be changed into a sawtooth form. This is the function of C_{ST} in

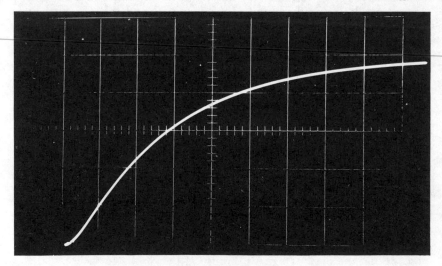

Fig. 3-13. Expansion of the leading edge of a waveform.

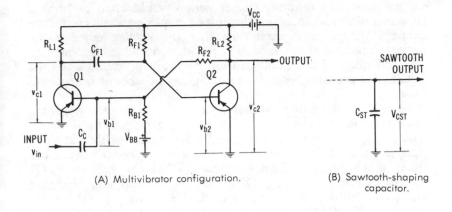

(A) Multivibrator configuration.

(B) Sawtooth-shaping capacitor.

(C) Operating waveforms.

Fig. 3-14. Basic monostable (one-shot) multivibrator.

Fig. 3-14B. Note that during the time that Q2 is cut off, C_{ST} charges through R_{L2} and forms one sawtooth excursion. Next, when Q2 suddenly goes into saturation again, C_{ST} discharges rapidly through Q2. The resulting sawtooth wave, V_{CST}, is in turn applied to the horizontal-deflection plates in the scope, and produces one excursion of the triggered-sweep action. It is evident that the sweep speed depends on the value of C_{ST} (time constant of the charging circuit). In practice, it is necessary to elaborate the basic time base somewhat. First, if an arbitrary vertical-signal waveform were used to trigger the multivibrator, the resulting trigger action would often be erratic. Therefore, a waveshaping section must be employed which produces a standard trigger pulse each time that the leading edge of the vertical-signal waveform appears. Second, control facilities must be provided whereby this standard trigger pulse is produced when the vertical-signal leading edge is positive-going or when it is negative-going.

Fig. 3-15 shows a basic waveshaping section that is used to trigger the monostable multivibrator. This waveshaping arrangement is called a Schmitt trigger circuit. It is essentially a bistable multivibrator. In other words, if Q1 is cut off, Q2 will be saturated, and the circuit will remain in this state until the base voltage on Q1 is changed to bring Q1 into conduction. Thereupon, Q2 is driven into cutoff, and the circuit will remain in this reversed state until the base voltage on Q1 is again changed to cut off Q1. In turn, Q2 simultaneously goes into saturation. This circuit action is a result of the common-emitter resistor, R_E, and the bias source, V_{EE}. Input and output waveforms are shown in Fig. 3-15C, for the case of a sine-wave input. Observe that the output is a square wave. No matter what the input waveform may be, the output will remain a square waveform. Thus, uniform triggering of the monostable multivibrator is ensured. Note that when the leading edge of the output waveform in Fig. 3-15B is applied to the input lead of the

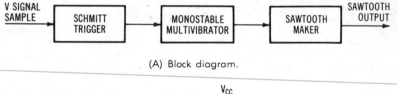

(A) Block diagram.

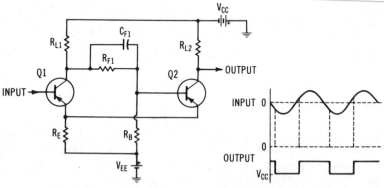

(B) Schmitt trigger configuration.

(C) Input and output waveforms.

Fig. 3-15. Monostable multivibrator preceded by a Schmitt trigger circuit.

56

multivibrator in Fig. 3-14A, the square waveform becomes differentiated by C_C and R_{B1}, thereby forming the trigger pulse, V_{in}, depicted in Fig. 3-14C.

We observe that triggering of the time base can occur only on the negative-going excursion of the input waveform in Fig. 3-15B. Of course, this is not always desirable—we might wish to display a positive-going waveform instead. Therefore, a triggered-sweep scope provides facilities for either positive or negative triggering. For this purpose, the Schmitt trigger in Fig. 3-15A is preceded by a phase inverter. A phase inverter effectively turns the input waveform "upside down" so that its negative-going leading edge is changed into a positive-going leading edge. Similarly, the positive-going leading edge of the input waveform is changed into a negative-going leading edge. This change permits switch control of negative-going triggering or positive-going triggering. Still other elaborations of the basic triggered time base are provided in modern scopes. These will be explained in greater detail in the next chapter.

COMMERCIAL DESIGN

At this point, it is instructive to consider a widely used commercial design of a solid-state triggered time base and horizontal-amplifier configuration, as shown in Fig. 3-16. The sweep generator circuits are part of the circuitry on the sweep generator and horizontal-amplifier circuit board. The Int-Ext switch (SW503) on the front panel determines whether the internal trigger signal or an external trigger signal will be used to start the sweep. In either case, the selected signal is coupled to the gate of transistor Q302. Internal triggering is accomplished by coupling a signal from the vertical amplifier, through transistor Q301 and dc level control R301, to trigger amplifier input stage Q302. Level control R301 is adjusted to obtain 0 volt dc at the gate of transistor Q302. The Level Set control varies the voltage on the source of transistor Q302 by changing the current level through transistor Q303. The source voltage of Q302 is set at 0 volt, when the Auto-Norm trigger switch (SW504) is in the Auto position. When the Auto-Norm switch is in the Norm position, the Trigger Level control on the front panel of the set performs the function of selecting the current through transistor Q303. In turn, it controls the point at which the sweep generator will trigger.

Both gain and dc level control are achieved as the signal is coupled through transistors Q304 and Q306, a differential amplifier. Capacitor C307 is a high-frequency ac coupler connected between the emitters of the differential amplifier. The output from this amplifier is coupled through emitter-follower output transistors Q305 and Q307 to switch SW505 (the +/− switch) on the front panel. Transistor Q308 is a

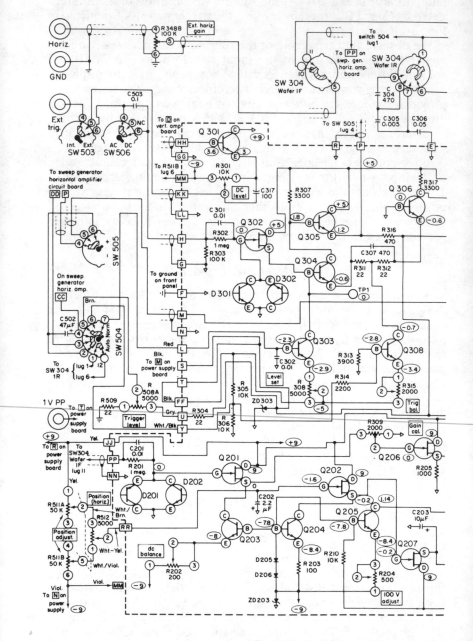

Fig. 3-16. A solid-state triggered time base

58

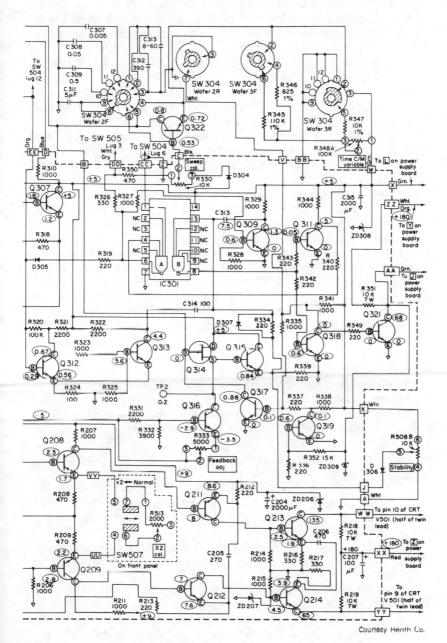

Courtesy Heath Co.

and horizontal-amplifier configuration.

constant-current source for transistors Q304 and Q306. The collector current of Q308 is determined by the Trig Bal control (R315) in its emitter circuit. The Trig Bal control is adjusted to present the proper dc level to the input of integrated circuit IC301. This integrated circuit is a dual Schmitt trigger. The "A" section of IC301 is used to shape the wave of the trigger signal, while the "B" section is used as a voltage sensor to turn the sweep off at the end of each sweep cycle.

With reference to Fig. 3-17, the output of IC301B (pin 8) is high most of the time. This positive voltage turns on transistor Q318 which, in turn, turns off Q317 and allows the selected sweep capacitor to charge. The high output from IC301B also turns on transistor Q311, which grounds one input pin (pin 4) of IC301A. This keeps other input pulses at pin 1 (IC301A) out of the circuit so that IC301B will not trigger before the sweep is completed. Stability control resistor R508B is set to bias the input of IC301B to a voltage level slightly more positive than the reset voltage. As the sweep capacitor charges, it overrides this voltage (after being coupled through source-follower and emitter-follower transistors Q314 and Q315), and continues to increase until the output of IC301B goes low. Then, transistor Q318 is turned off and transistor Q317 is turned on, shorting out the sweep capacitor. Transistor Q311 is also turned off, which causes pin 4 of IC301A to go high.

The next positive-going pulse to pin 1 drives the output low, turning off transistor Q309. A positive pulse is then coupled through capacitor C313 and sets the output of IC301B low. After the input pulse at IC301A is gone, pin 1 again is low and pin 6 (IC301A) goes high and turns on transistor Q309. The negative pulse coupled through capacitor C313 causes the output of IC301B to go high. The output stays high because this input is biased between the turn-on and turn-off points of the Schmitt trigger. Then, the process repeats itself. Transistor Q321 is the unblanking amplifier (Fig. 3-16). This amplifier receives signals from the sweep circuits to properly bias the crt, turning the electron beam on and off as required. Feedback from Q321, through transistor Q319, holds the start of the trace until Q321 has returned to its quiescent condition from the preceding sweep. This prevents the shortening of the trace due to an insufficient rise time in transistor Q321.

Transistors Q312 and Q313 form a Schmitt trigger circuit that is modified by a negative-feedback signal coupled from the collector of Q312 back to its base through resistor R320. The base of Q312 is capacitively coupled to ground to form a free-running multivibrator. This multivibrator operates on the three high ranges of the Time C/M switch and is used to provide a better base line on the crt. Normally, when a triggered sweep circuit is operating in the automatic mode, the base line will fade as the sweep rate is increased. To prevent this

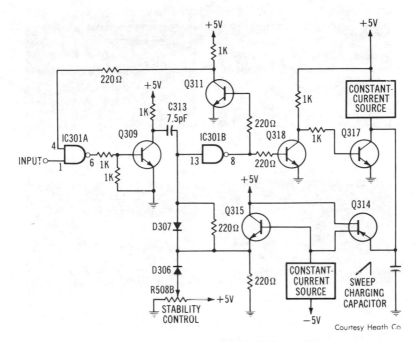

Courtesy Heath Co.

Fig. 3-17. Schmitt trigger circuitry.

fading, the multivibrator is used to provide an increasing trigger repetition rate as the sweep speed is increased. At the lower sweep speeds, the generator will be automatically "retriggered" at about 50 Hz. Should a higher frequency be available, such as that from the multivibrator described above, the generator circuit will lock on to the faster rate.

The Time C/M switch determines the value of the sweep capacitor, and the amount of current flowing through transistor Q322. As the sweep capacitor charges, a positive-going ramp voltage (sawtooth) is generated. The speed of the horizontal sweep is determined by the particular timing capacitor that was chosen, and by its charging current. The effect of horizontal sweep speed on the display of a narrow pulse is exemplified in Fig. 3-18. At a sweep speed of 0.02 ms/cm, the pulse appears to have zero rise time and zero fall time; it appears to have sharp corners without any evidence of rounding. As the sweep speed is increased to 0.2 μs/cm, it becomes evident that the pulse has a finite rise time, and that it has rounded corners. When the sweep speed is further increased to 0.04 μs/cm, it is seen that the rise time is approximately 0.02 μs. In this high-speed display, most of the top portion of the pulse is off-screen to the right.

61

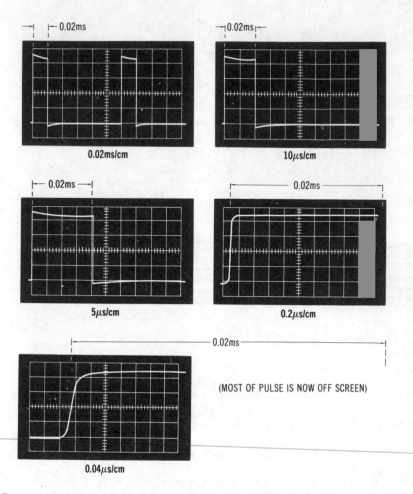

Figure 3-18. Progressive expansion of a 20-microsecond pulse on the screen of a triggered-sweep scope.

Rise time is measured from the 10% point to the 90% point along the leading edge, as shown in Fig. 3-19. This definition avoids the confusion that could be caused by various modes of cornering. Similarly, *fall time* is measured from the 90% point to the 10% point along the trailing edge of the waveform. Precise measurement of rise time (or fall time) requires that the leading or trailing edges of a fast-rise pulse be considerably expanded by being displayed on a high-speed sweep. Note that this type of waveform has invisible leading and trailing edges when displayed on a slow-speed sweep. In other words, the writing rate of the leading and trailing edges is much

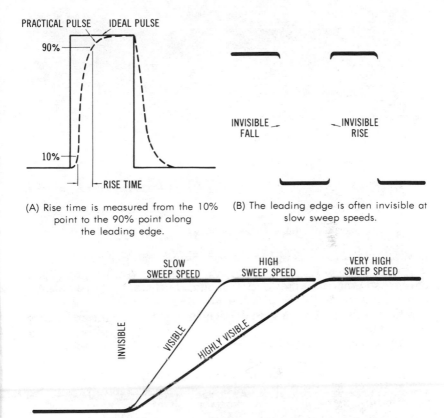

(A) Rise time is measured from the 10% point to the 90% point along the leading edge.

(B) The leading edge is often invisible at slow sweep speeds.

(C) Visibility of the leading edge improves with waveform expansion.

Fig. 3-19. Basic principles of a triggered-sweep display.

faster than the writing rate along the flat top when displayed on a slow-speed sweep. On the other hand, when the leading edge is greatly expanded on a high-speed sweep, the writing rate is more nearly uniform over the entire pattern. Note in passing that as the sweep speed is progressively increased, the intensity control of the scope must be advanced to maintain normal pattern brightness.

Synchronization

PRINCIPLES OF SYNCHRONIZATION

Synchronization is defined as the precise matching of two waves or functions. In other words, two electrical events are synchronized when they are so timed that they occur simultaneously, or in step with each other. In the complete *absence of synchronization,* a displayed waveform will drift or "run" to the right or to the left on the crt screen. In a *marginal sync operation,* a displayed waveform will be locked for a time on the screen, but will then "break sync" and slide horizontally to the left or right. As exemplified in Fig. 4-1, marginal sync lock may result in "jitter" of the waveform. In the majority of cases, the technician will be observing a signal that has some regularly recurring peak or dip in its amplitude. In other words, it is made up of cycles that repeat regularly and can, therefore, be synchronized with the trace of the oscilloscope as the trace sweeps across the oscilloscope screen.

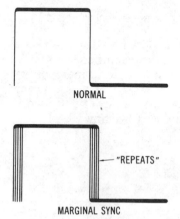

NORMAL

"REPEATS"

MARGINAL SYNC

Fig. 4-1. Pattern "jitter" may be caused by marginal synchronization.

When an oscilloscope has been properly synchronized with a signal, the signal will appear to be stationary on the screen, and one or more cycles of the signal can be seen. If the oscilloscope is slightly out of synchronization, the waveform will appear to move slowly across the screen, either to the right or to the left. The waveform can be made to "stand still" even without a sync signal if the operator carefully adjusts the fine frequency control. It will not remain stationary very long, though, because of the tendency of the sweep oscillator to wander in frequency. The situation is quite similar to that of a television receiver that has lost the sync signal. The receiver operator varies the frequency of the sweep oscillator by adjusting the hold control, and when the oscillator frequency agrees with the sweep frequency of the tv signal, the picture is held stationary on the screen, but only as long as the operator is willing to keep adjusting the hold control.

Certain types of signal will be more difficult to synchronize than others. These include signals having little information of a recurring or repeating nature, and signals having few pronounced peaks or dips. As the frequencies of both the signal being viewed and the oscilloscope sweep are increased, synchronization also becomes more difficult. The reason for these difficulties will become more apparent later when we consider the process of synchronization. As an example, the video signal output from the picture detector in a tv receiver is often difficult to synchronize when displayed on a scope with a free-running sweep. *However, if the external sync function of the scope is used to synchronize the sweep from the vertical integrator output, the pattern can be tightly locked on the screen.* This is because the vertical integrator removes all of the horizontal sync pulses and equalizing pulses, leaving only the vertical sync pulse (Fig. 4-2).

If a signal does not repeat at regular intervals, it is called a transient or nonrecurrent waveform. Even if a signal repeats at regular intervals, it may approximate a transient in case there is a long waiting interval between the repetitions. Sometimes there may be many waveforms of no interest that occur between the repetitions of the desired waveform. As an illustration, a vertical interval test signal is transmitted in each vertical-blanking interval of the composite color-tv signal (Fig. 4-3). The foregoing waveforms cannot be displayed with a scope that has a free-running sweep. On the other hand, such waveforms can be displayed with a scope that has triggered sweeps. It is sometimes advantageous to use a long-persistence crt, particularly when the signal is of the "one-shot" type and is not repeated at any time. Note that highly sophisticated lab-type scopes that have a screen-pattern storage function are also available. In other words, a "one-shot" transient can be maintained visibly on the screen for hours, if desired.

Various general-purpose scopes provide a choice of recurrent-

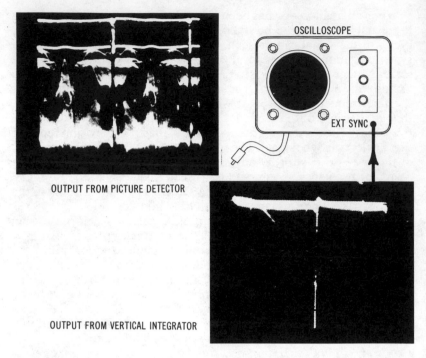

OSCILLOSCOPE

EXT SYNC

OUTPUT FROM PICTURE DETECTOR

OUTPUT FROM VERTICAL INTEGRATOR

Fig. 4-2. External sync function provides a stable lock of the composite video signal.

sweep mode and trigger-sweep mode of operation. This is called a dual-mode sweep function. A scope with this feature is pictured in Fig. 4-4. In its recurrent-sweep mode, it is a conventional continuous-sweep type of oscilloscope. The sweep oscillator is adjustable up to 3.58 MHz, permitting lock-in of high-frequency signals. The sweep ranges are marked not only with frequency indications, but also with time-base references (milliseconds or microseconds per centimeter). Preset tv V and H sweep positions are provided for convenience in tv servicing. A 60-Hz sine-wave horizontal-deflection function is also provided, with adjustable phase, for use in sweep-alignment procedures. This topic is explained further in a following chapter.

Automatic sync is a simplified form of triggered sweep; the scope illustrated in Fig. 4-4 provides automatic sync capability. This mode of operation displays a horizontal base line on the crt screen, whether there is a vertical input signal or not. When a vertical input signal is applied, the pattern automatically locks in sync, as shown in the diagram of Fig. 4-5. Although a choice of positive or negative sync is available, there is no control of the trigger level, as provided by a scope that has complete triggered-sweep capabilities.

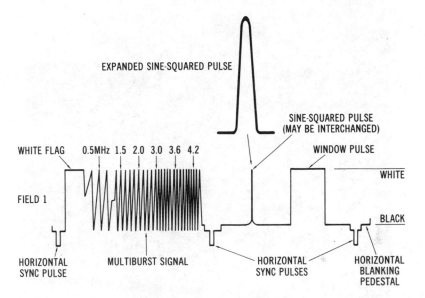

EXPANDED SINE-SQUARED PULSE

SINE-SQUARED PULSE
(MAY BE INTERCHANGED)

WHITE FLAG   0.5MHz 1.5 2.0 3.0 3.6 4.2

WINDOW PULSE

WHITE

FIELD 1

BLACK

HORIZONTAL
SYNC PULSE

MULTIBURST SIGNAL

HORIZONTAL
SYNC PULSES

HORIZONTAL
BLANKING
PEDESTAL

(A) First VITS scanning line.

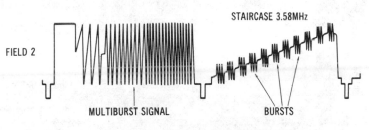

STAIRCASE 3.58MHz

FIELD 2

MULTIBURST SIGNAL

BURSTS

(B) Second VITS scanning line.

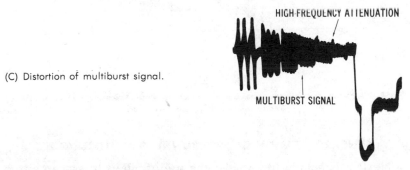

HIGH-FREQUENCY ATTENUATION

(C) Distortion of multiburst signal.

MULTIBURST SIGNAL

Fig. 4-3. Vertical Interval test signal. (These waveforms are transmitted by color-tv stations on two lines of the vertical retrace interval.)

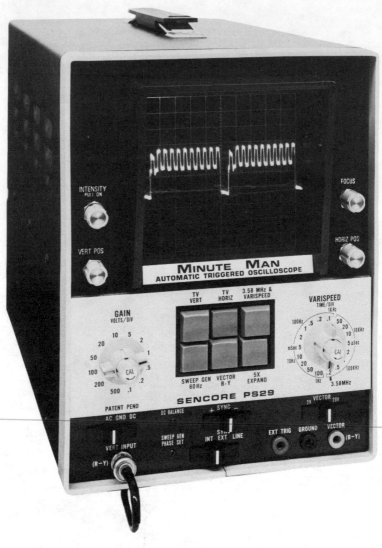

Fig. 4-4. A tv-service oscillosope that permits an operational choice of the recurrent-sweep mode or the trigger-sweep mode.

SYNCHRONIZATION OF FREE-RUNNING MULTIVIBRATORS

Next, it is helpful to consider the use of pulses to synchronize a transistor multivibrator in a circuit. Fig. 4-6 shows the effect of positive pulses on the multivibrator base waveform. Note that when a positive

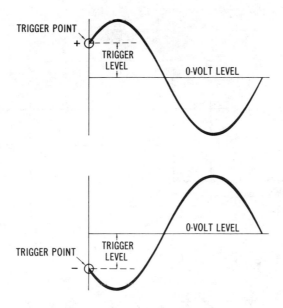

Fig. 4-5. Automatic sync action provides a choice of positive or negative triggering; however, no control of trigger level is available.

pulse of insufficient amplitude to drive the base above cutoff (zero volts) is applied to a nonconducting transistor at instant A, it does not cause switching action. In other words, its only effect is to reduce the negative bias slightly, as shown at A´. Note also that when a positive pulse is applied to a transistor that is already conducting (points B or C), the pulse serves merely to increase the base voltage momentarily. That is, the pulse has no effect on multivibrator action at this instant. Observe that with the exceptions of the variations in the v_b waveform at times A´, B´, and C´, the multivibrator is essentially free-running. However, if the positive pulse is applied at instant D, it has sufficient amplitude to overcome the negative voltage on the base of Q1, and it drives the base above zero volts. Accordingly, the period of the multivibrator is shortened by the interval, EF.

We will recognize that for proper synchronization, the natural period of the multivibrator in its free-running state must be greater than the time interval between sync pulses. Under this condition, the positive trigger pulses cause the switching action to occur earlier in the cycle than it would occur in the free-running state. Hence, the transistor conducts at E, G, and G´ in Fig. 4-6. Otherwise, in the absence of sync pulses, the transistor would have conducted later in each instance. In this manner, synchronizing action forces the multivibrator frequency to become the same as the repetition rate of

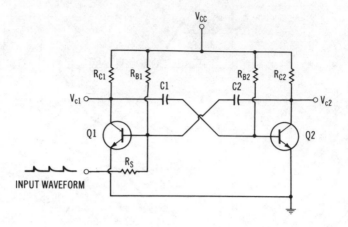

(A) Circuit configuration.

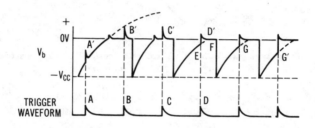

(B) Base waveform with positive sync pulses.

Fig. 4-6. Synchronization of a transistor multivibrator.

the sync pulses. Of course, the multivibrator may be synchronized to a submultiple of the trigger frequency, if both frequencies are such that every second, third, fourth, etc., sync pulse occurs at the correct time to drive the base voltage of the nonconducting transistor above cutoff.

TRIGGERED-SWEEP SYNCHRONIZATION

With reference to Fig. 4-7, synchronization of a triggered-sweep oscilloscope occurs incidentally with adjustment of the time-base controls. The example diagram shown provides a choice of internal, external, line (60 Hz), TVH (7875 Hz), or TVV (30 Hz) sync. The sync inverter permits triggering on either the leading edge or on the trailing edge of a waveform. Regardless of the type of sync that is utilized, the sync signal is first processed by a trigger shaper which develops a standard sync trigger pulse. This trigger pulse is then applied to the

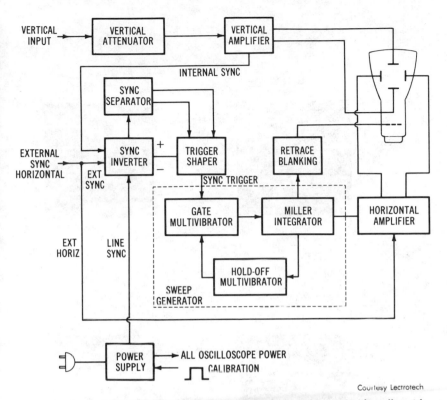

Fig. 4-7. Synchronization in a triggered-sweep oscilloscope occurs incidentally with adjustment of the time-base controls.

sweep-generator section, where it will be admitted by the gate multivibrator, provided that the hold-off multivibrator has sensed the completion of a sawtooth sweep waveform. In turn, the gate multivibrator drives a Miller integrator, which generates a highly linear ramp (sawtooth wave). The gate multivibrator may be synchronized by the leading edge of the waveform, by its trailing edge, or by some intermediate rising or falling interval within the waveform.

Typical high and low trigger points on the leading and trailing edges of a trapezoidal waveform are depicted in Fig. 4-8. It is also possible to trigger the sweep circuit at any point along the flat top of the waveform. However, the internal-sync function of the scope cannot be used in this case. Instead, an external-sync setting such as shown in Fig. 4-4 must be utilized. In turn, a suitable sync signal is applied to the sync-input terminals. This sync signal is typically obtained from a pulse generator which, in turn, is locked to the vertical-input signal. Effectively, a delayed sync pulse is applied to the sync phase splitter in

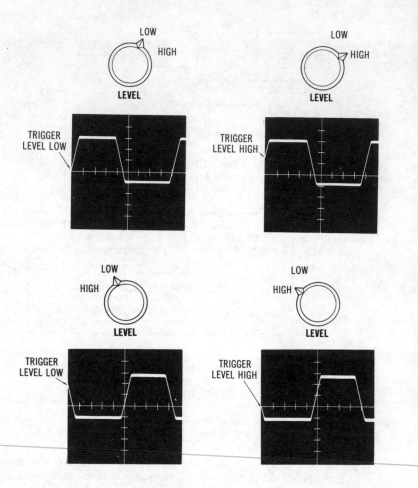

Fig. 4-8. The starting point of the displayed waveform can be chosen at any desired point on the leading or trailing edge by an adjustment of the level control.

order to trigger the sweep circuit at some point along the flat top of the waveform being displayed. The exact point at which the sweep circuit is triggered depends on the repetition rate to which the pulse generator is set (Fig. 4-9).

OVERSYNCHRONIZATION

Both types of sweep oscillators are subject to oversynchronization if too much sync signal is applied, and the effects on the resulting waveforms are similar. Figs. 4-10A and 4-10B show waveforms that

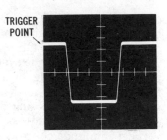

Fig. 4-9. A delayed sweep is required to trigger the time base at some point along the flat top of the waveform.

were photographed at different points in a multivibrator sweep circuit of an oscilloscope. Fig. 4-10A shows the signal obtained at the plate of the first section, and Fig. 4-10B, the signal at the output of the discharge section. Fig. 4-10C illustrates the waveform that was actually displayed on the screen of the oscilloscope. Fig. 4-10A shows that one cycle of sweep contains parts of three cycles of sine-wave signal and that the other cycle of sweep contains parts of two cycles of signal. Fig. 4-10B depicts that the sweeps travel at a constant rate but that alternate cycles are of different lengths. Each cycle of sweep and retrace can be seen to correspond to certain portions of the waveform in Fig. 4-10C.

If oversynchronization is carried to extremes, the sawtooth waveform of Fig. 4-10B may even degenerate into one large cycle followed by two or three very small ones. In such a case, the waveform

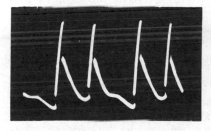

(A) Signal at the plate of first section.

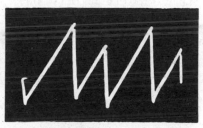

(B) Sawtooth signal developed at second section.

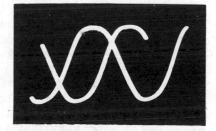

(C) Waveform seen on oscilloscope screen.

Fig. 4-10. Oversynchronization of a multivibrator sweep circuit.

on the screen would also be greatly distorted. Some manufacturers have designed circuits to lessen or eliminate the possibility of oversynchronization. One method is to use a limiter stage ahead of the point of injection of the sync signal. In oscilloscopes that do not have provision for limiting oversynchronization effects, it is sometimes found that a change in the vertical amplitude setting will affect the sweep operation, causing the waveform to fall either into or out of sync. These are usually cases where the sync takeoff point follows the vertical amplitude control. Therefore, when the vertical amplitude is changed, the amplitude of the sync signal changes with it, with the result that synchronization may be affected.

For simplicity, the sync signals shown in the preceding examples have been sine-wave signals. In cases where the sync signal is taken as a part of the signal in the vertical amplifiers of the oscilloscope, it can take on any form, depending on the waveform being viewed, as shown in Fig. 4-11. Generally, stable synchronization will be more easily attained with signals of a peaked or sharply pulsed nature rather than with those of a more even nature. Sometimes a signal may have more peaks at its negative extreme than at its positive extreme, or vice versa. A sync polarity-reversal switch on the oscilloscope may help the operator obtain stable synchronization in these cases (see Fig. 4-12). If such a switch is not included, the same effect can sometimes be obtained by taking the signal to be observed from a point where the signal is 180 degrees out of phase with respect to the signal at the first point. The signals between two successive stages in an amplifier usually have this phase reversal.

Fig. 4-13 illustrates the panel of a scope showing a typical group of trigger controls. When triggered sweep is used, oversynchronization results if the stability control is improperly adjusted. With reference to Fig. 4-14, a typical triggered-sweep scope displays only a horizontal line on the screen when the stability control is set to the extreme right-hand end of its range, even though a vertical-input signal is applied. Next, when the stability control is set to the midpoint of its range, the vertical-input signal is displayed on the screen. Finally, when the stability control is set to the extreme left-hand end of its range, the screen is blank. These extreme examples correspond to the waveform distortion that occurs when a scope with free-running sweep is oversynchronized. Note that oversynchronization is impossible for a scope that has automatic sync action. However, as shown in Fig. 4-15, although any vertical-input signal will be automatically synchronized, the horizontal-sweep speed may need an adjustment in order to get the desired pattern aspect. A waveform may be greatly expanded on the screen when automatic sync is used, just as if complete triggered sweep were being utilized—the only distinction is that the operator has no choice of sync level when automatic sync is used.

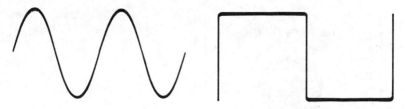

(A) A sine wave is the basic steady-state waveform.

(B) A square wave is the basic transient-state waveform.

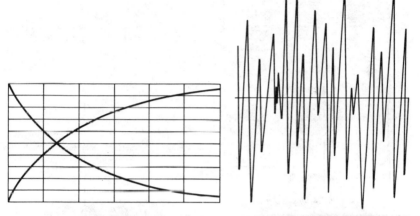

(C) An exponential wave is the basic growth/decay waveform.

(D) A noise wave is the basic random waveform.

Fig. 4-11. The four fundamental waveforms.

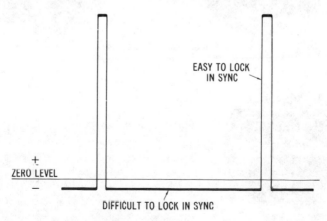

EASY TO LOCK IN SYNC

+
ZERO LEVEL
−

DIFFICULT TO LOCK IN SYNC

Fig. 4-12. A pulse waveform that has a very small negative excursion.

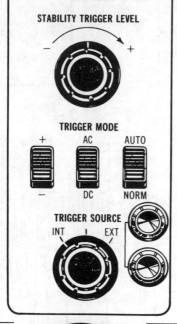

STABILITY TRIGGER LEVEL

– +

TRIGGER MODE

+ AC AUTO

– DC NORM

TRIGGER SOURCE

INT EXT

Fig. 4-13. Panel of a scope showing a typical group of trigger controls.

(A) Set to the extreme right-hand end of its range.

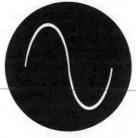

(B) Set to the midpoint of its range.

(C) Set to the extreme left-hand end of its range.

Fig. 4-14. The stability trigger-level control.

(A) Horizontal base line is displayed in absence of signal input.

(B) Pattern is displayed and automatically locked in sync when input signal is applied.

(C) Horizontal sweep speed may need adjustment.

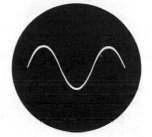

Fig. 4-15. Response of a scope with an automatic sync function.

CHAPTER 5

Vertical Amplifiers

VERTICAL-AMPLIFIER CHARACTERISTICS

Almost all present-day oscilloscopes contain vertical amplifiers in order to increase the signal amplitude before it is applied to the deflection plates of the crt. Many scopes also have provision for making a direct connection to the vertical (and horizontal) deflection plates. However, a signal of comparatively high amplitude is required to obtain a useful amount of deflection when making a direct connection to the deflection plates. The more elaborate lab-type scopes may offer interchangeable plug-in vertical (and horizontal) amplifiers. Thereby, the operator can use the same mainframe with a general-purpose vertical amplifier, a specialty measurement vertical amplifier, or other type of amplifier. As a general rule, narrow-band amplifiers can be designed to have high gain, whereas, wide-band amplifiers are comparatively limited in gain. Although an extremely wide-band amplifier can be constructed to have very high gain, *noise* in the output makes the amplifier impractical. Therefore, a rule-of-thumb gain/bandwidth product must be taken into account by scope designers.

One model of service-type scope is rated for a deflection sensitivity of 15 volts rms per inch at the vertical-deflection plates, and for a deflection sensitivity of 15 mV per inch at the vertical-amplifier input. Thus, in this example, the vertical amplifier provides a voltage gain of 1000. The crt has frequency capability up to 50 MHz, whereas, the vertical amplifier, in this example, has frequency response to 5 MHz. Service-type scopes are rated for vertical-amplifier sensitivity on the basis of rms volts-per-inch, of peak-to-peak volts-per-inch, of rms volts-per-centimeter, of peak-to-peak volts-per-centimeter, of rms volts-per-division, or of peak-to-peak volts-per-division. In the latter rating, the length of a screen division is established by the manufacturer. The relations (Fig. 5-1) of sine-wave rms, peak-to-peak, and peak voltages are as follows:

> Peak-to-peak voltage = 2 × peak voltage
> Peak voltage = ½ × peak-to-peak voltage
> Rms voltage = 0.707 × peak voltage
> Peak voltage = 1.414 × rms voltage
> Peak-to-peak voltage = 2.83 × rms voltage

Furthermore, if you are using a dc scope, you need to know that when you apply a 1.5-volt battery across the vertical-input terminals of a dc scope, the trace will move the same vertical height as when a 1.5 peak-to-peak sine-wave voltage is applied.

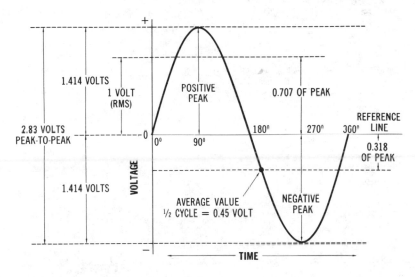

Fig. 5-1. Voltage relations in a pure sine wave.

One inch equals 2.54 centimeters; thus, a scope having a sensitivity of 0.01 rms volt-per-inch has an equivalent sensitivity of approximately 0.004 rms volt-per-cm. Note that all of the following ratings are exactly equal to one another:

Rms volts per inch	0.010
Peak-to-peak (or dc) volts per inch	0.028
Rms volts per centimeter	0.004
Peak-to-peak (or dc) volts per centimeter	0.011

SAMPLING OSCILLOSCOPES

Some lab-type oscilloscopes have vertical amplifiers, with a very considerable effective bandwidth. For example, one design provides response up to 3900 MHz. To obtain this extreme bandwidth, a sampling technique is used, as pictured in Fig. 5-2A. Conventional oscilloscopes are limited in bandwidth to frequencies in the megahertz region. However, sampling oscilloscopes have response into the gigahertz region. This design samples the input waveform and then reconstructs the waveform for display from the samples taken during many recurrences of the input waveform (Fig. 5-2B). Therefore, a sampling oscilloscope cannot be used to display one-shot transients. To reconstruct a waveform, the sampling pulse "turns on" the sampling circuit for a brief instant, and the amplitude at this point is displayed by the electron beam in the crt. After a brief pause, the following cycle of the input waveform is again sampled, and the amplitude at this next point is displayed by the crt beam which has moved slightly to the right in the meantime. As many as 1000 samples

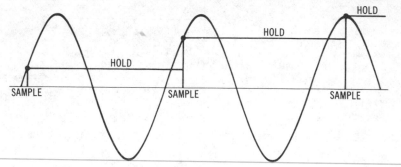

(A) Reconstructed waveform.

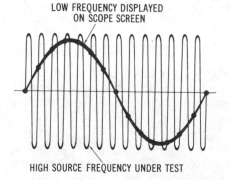

(B) Waveform being sampled.

Fig. 5-2. Sample-and-hold process permits analysis of high-speed events.

may be processed to reconstruct the input waveform for display on the crt.

TYPICAL VERTICAL-AMPLIFIER ARRANGEMENTS

A vertical-amplifier configuration for a service-type scope with a sensitivity of 30 mV p-p/cm and a frequency response to 5 MHz at the −3-dB point is shown in Fig. 5-3. The input signal to the vertical amplifier is coupled through resistor R1 and capacitor C1 to the gate of FET Q1. Resistor R1 serves as a protective resistor for transistor Q1 in the event that an excessive input signal voltage is accidentally applied. Additional overload protection is provided by diodes D1 and D2; these are transistors connected to simulate zener diodes in order to limit the Q1 gate voltage to ±9 volts. Capacitor C1 provides improved high-frequency response across resistor R1. Transistor Q1 operates as a source follower; it provides high input impedance and moderate output impedance. Note that transistor Q2 serves as a constant-current source for Q1 and, thereby, stabilizes its operation. Diodes D4 and D5 provide a practically constant-voltage source (total of 1.2 volts) for the base of transistor Q2. In turn, the current through resistor R2 is virtually constant; R2 is adjusted to set the source voltage of Q1 at zero volt under no-signal conditions.

When an input signal is applied to the gate of transistor Q1, the source current remains constant, and only the source voltage changes. This signal-voltage variation is applied to the gain control R404, and thence to the gate of the source-follower transistor Q3. Note that transistor Q4 provides a constant-current supply for transistors Q5 and Q6. This constant-current circuit serves as a common-emitter load for transistors Q5 and Q6, thereby establishing a stable operating point for the two transistors, and for the following stages. Next, the signal output from source-follower transistor Q3 is amplified by Q5. It is evident that a portion of the signal at the base of transistor Q5 appears also at the emitter of Q5; because transistors Q5 and Q6 have a common emitter load, the signal at the emitter of Q5 is coupled to the emitter of Q6. In other words, the single-ended input signal is changed into a double-ended (push-pull) signal.

Observe that transistor Q6 in Fig. 5-3 operates as a common-base amplifier; its base voltage is determined by the setting of potentiometer R406, the vertical-position control. This vertical-position control moves the crt beam vertically by applying an adjustable potential to the base of transistor Q6, thereby producing a certain amount of dc unbalance in the following vertical-amplifier system. Note that when the collector voltage of transistor Q5 decreases, its emitter voltage increases. As a result, the forward bias on transistor Q6 decreases and its collector voltage increases. Since the signal at the

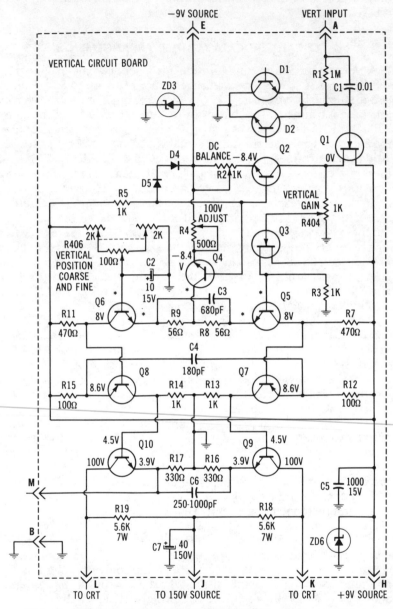

*This voltage depends on operating parameters of transistor Q3.

Courtesy Heath Co.

Fig. 5-3. A typical vertical amplifier configuration.

collector of Q6 is 180° out of phase with the signal at the collector of transistor Q5, push-pull amplification is obtained. Note that C3 is an emitter *partial-bypass* capacitor. It increases emitter degeneration at low frequencies, or, it provides a relative increase of high-frequency output. Therefore, the vertical-amplifier frequency response is maintained more nearly uniform. Resistors R8 and R9 determine the dc gain of the stage. The driver transistors Q7 and Q8 operate in the common-emitter mode. Transistors Q7 and Q8 not only provide gain, but also buffer transistors Q5 and Q6 from the changing current demand of the output transistors. Capacitor C4 is an emitter partial-bypass capacitor that provides a relative increase of high-frequency output. Final amplification is provided by transistors Q9 and Q10 for energizing the vertical-deflection plates in the crt.

Some vertical-amplifier configurations use peaking coils to assist in obtaining an extended high-frequency response. Fig. 5-4 shows basic

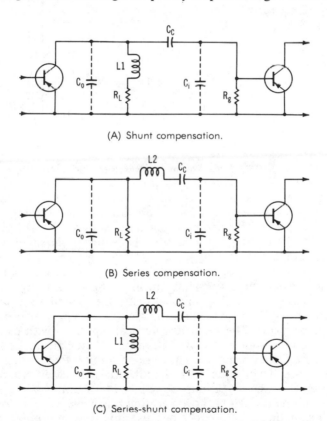

(A) Shunt compensation.

(B) Series compensation.

(C) Series-shunt compensation.

Fig. 5-4. Basic peaking-coil arrangements.

shunt, series, and series-shunt compensation circuits for transistor stages. The residual circuit capacitances are indicated at C_0 (output capacitance) and C_i (input capacitance). At high frequencies, L_1 becomes parallel-resonant with C_0 and C_i, thereby increasing the effective value of the collector load impedance and, in turn, increasing the stage gain. Again, at high frequencies, L_2 becomes series-resonant with C_0 and C_1, resulting in a voltage magnification across L_2 equal to its Q value. Fig. 5-5 shows the frequency response curves for various values of peaking-coil inductances.

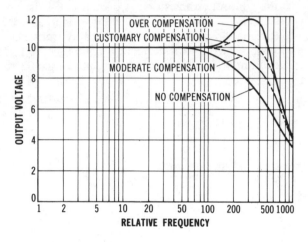

Fig. 5-5. Frequency response curves.

Vertical amplifiers, in lab-type scopes with extended high-frequency response, often utilize a distributed-amplifier configuration, such as shown in Fig. 5-6. This arrangement uses many very small peaking coils in up to 30 stages of amplification. The effect of the small peaking coils is to compensate for the grid and plate capacitances of the tubes. In effect, the amplifier operates as an artificial transmission line, with gain provided at each section along the line. The load resistors R_0 have very low values, such as 100 ohms. A distributed amplifier is complex and costly, but it can provide uniform high-frequency response out to 50 MHz, or more. The gain of a vertical amplifier with extended high-frequency response tends to be less than that of an amplifier with restricted frequency response. The reason for this is that random noise becomes more of a problem as the bandwidth is increased. Therefore, the sensitivity of the scope must be reduced to maintain a reasonable signal-to-noise ratio. Note, however, that this consideration does not apply to sampling oscilloscopes because this type of scope employs narrow-band amplification following the input sampler.

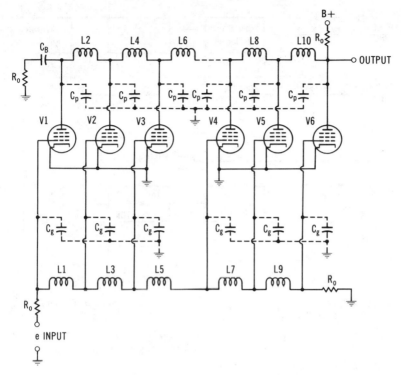

Fig. 5-6. Basic distributed-amplifier configuration.

Delay Lines

Triggered-sweep oscilloscopes with high-speed time bases often include a delay line in the vertical-amplifier section, as shown in the block diagram of Fig. 5-7. The delay line permits the time base to "get started" before the vertical-input signal arrives at the crt deflection plates. Referring to Fig. 5-8, it can be seen that the leading portion of a fast-rise pulse will be "lost" unless a delay line is used. A delay line provides a "lead time" of a fraction of a microsecond to allow the time base to respond to the sync trigger. Delay lines are essentially artificial transmission lines, as shown in Fig. 5-9. The propagation time of an artificial line provides the needed delay from input to output. An elaborate delay line typically includes several dozen sections; the capacitance values throughout the line are quite critical and are precisely adjusted at the time of manufacture.

Cutoff Characteristic

When peaking coils are used in a vertical-amplifier arrangement, the cutoff characteristic is comparatively steep, as shown in Fig. 5-10. In

85

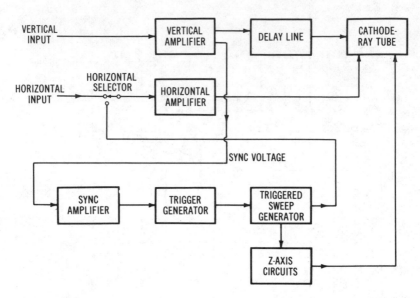

Fig. 5-7. Laboratory scopes include a delay line in the vertical channel.

other words, peaking coils develop a rapid high-frequency rolloff. Vertical amplifiers that do not employ peaking coils, such as the configuration shown in Fig. 5-3, have a comparatively gradual cutoff characteristic; the high-frequency rolloff is slow. Note that peaking coils are used to obtain a uniform response out to a higher frequency than would be obtained otherwise. The chief disadvantage of peaking coils and their associated steep cutoff characteristic is depicted in Figs. 5-10B and 5-10C. In other words, when an input waveform has

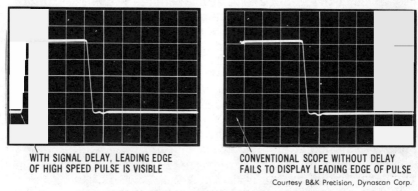

WITH SIGNAL DELAY, LEADING EDGE
OF HIGH SPEED PULSE IS VISIBLE

CONVENTIONAL SCOPE WITHOUT DELAY
FAILS TO DISPLAY LEADING EDGE OF PULSE

Courtesy B&K Precision, Dynascan Corp.

Fig. 5-8. The entire leading edge of a fast-rise pulse may be lost unless a vertical delay line is utilized.

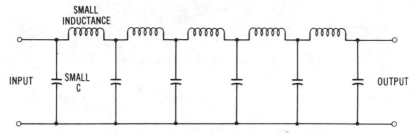

Fig. 5-9. A basic delay line configuration.

harmonic frequencies past the vertical-amplifier cutoff frequency, a steep cutoff characteristic will cause the displayed pattern to exhibit overshoot and ringing. This distortion can be reduced by using a vertical amplifier with a slower high-frequency rolloff. Sophisticated oscilloscopes are generally designed to have a Gaussian response, as shown in Fig. 5-11. This is the optimum rolloff characteristic; its rate is 10 dB per octave (from f to 2f, such as 10 MHz to 20 MHz).

VERTICAL STEP ATTENUATORS

To accommodate a wide range of vertical input-signal voltages and to avoid vertical-amplifier overload, a vertical step attenuator is provided, as diagrammed in Fig. 5-12. To "fill in" between steps, a vernier or fine vertical-gain control, such as potentiometer R404 in Fig. 5-3, is also provided. However, a vertical step attenuator must provide a high input impedance, such as 1 megohm. Therefore, a simple resistive-type voltage divider is inadequate; its stray capacitances and associated capacitances would produce severe high-frequency distortion. Therefore, *frequency-compensated* step attenuators are utilized. An eight-step configuration is shown in Fig. 5-14B. It provides a nominal input resistance of 1 megohm on each step. Note that each of the input resistors (R402, R404, R406, and so on) is shunted by a trimmer capacitor. The technical consideration for this is that the time constant of each input section can be adjusted to equal the time constant of the corresponding output section. When this is done, complex waveforms will pass through the attenuator without distortion.

The practical effect of a trimmer adjustment in a vertical step attenuator is shown in Fig. 5-13. If a 10-kHz square wave, for example, is applied to the scope, it will be displayed in a distorted form on the crt screen unless the compensating trimmer adjustments are correct. Once the trimmer capacitors have been correctly adjusted, they will seldom require attention. A step attenuator provides a constant vertical-input resistance, such as 1 megohm, on each step. It also

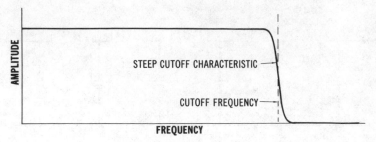

(A) Frequency response curve for a vertical amplifier with rapid high-frequency rolloff.

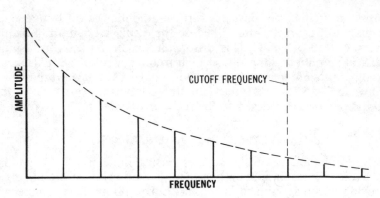

(B) Input square wave has harmonic frequencies past the vertical amplifier cutoff frequency.

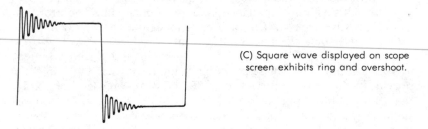

(C) Square wave displayed on scope screen exhibits ring and overshoot.

Fig. 5-10. Development of the cause of overshoot and ringing.

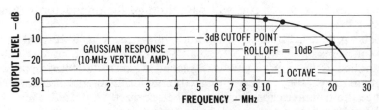

Fig. 5-11. Sophisticated oscilloscopes generally have a high-frequency rolloff of 10 dB per octave (Gaussian response).

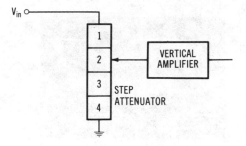

(A) Direct-coupled vertical input.

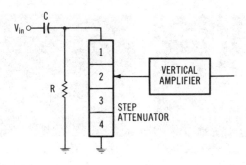

(B) RC-coupled vertical input.

Fig. 5-12. Oscilloscope input arrangements.

provides a constant vertical-input capacitance, such as 25 pF, on each step. In turn, the input impedance at very low frequencies and at dc level is 1 megohm. On the other hand, the input impedance to the attenuator at high frequencies, such as 4 MHz, is approximately 1500 ohms. This perhaps surprising decrease of input impedance at high frequencies is due to the decreasing reactance of the 25-pF input capacitance as the input frequency is increased.

Note that the vertical step attenuator, in Fig. 5-14B, is preceded by a series capacitor (C403), which can be shorted out by switch S401. This blocking capacitor changes the vertical-amplifier response from a dc to an ac response. When the scope is operated on its ac-input function, any dc component that might be present in the applied signal voltage is rejected. This process is exemplified in Fig. 5-15. The output from a half-wave rectifier is applied to the vertical-input terminal of the scope. The zero-volt level on the crt screen is determined by the resting level of the horizontal trace when no input signal is applied. In turn, when the waveform shown in Fig. 5-15C is applied to the vertical-input terminal of the scope, the zero-volt level on the crt screen will cut

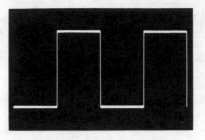

(A) Undistorted input waveform.

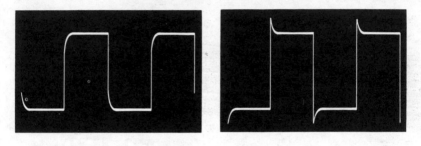

(B) Undercompensated.   (C) Overcompensated.

Fig 5-13. Distortion of a 10-kHz square wave by an improperly adjusted vertical step attenuator.

through the pattern at its average value, if blocking capacitor C is in the circuit. On the other hand, if capacitor C is shorted out, the zero-volt level on the crt screen will then coincide with the bottom of the pattern. In other words, because the vertical-input signal has a dc component (average value of the waveform), the pattern will rise or fall on the crt screen, depending upon whether capacitor C is in the circuit or out of the circuit.

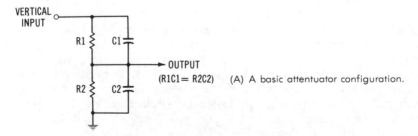

(A) A basic attentuator configuration.

Fig. 5-14. A frequency-compensated vertical

90

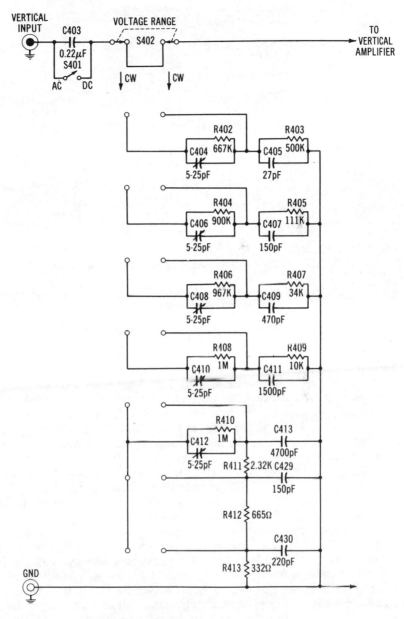

(B) A commercial attenuator arrangement.

step-attenuator configuration.

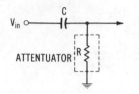

(A) An ac-input (RC-coupling) path.

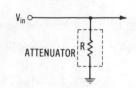

(B) A dc-input (direct-coupling) path.

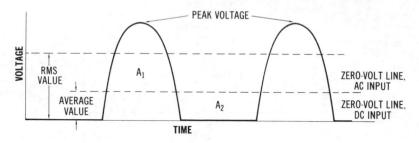

(C) The zero-volt axis with respect to the displayed waveform.

Fig. 5-15. An ac-input coupling and a dc-input coupling circuit for an oscilloscope and the resulting waveform display.

VERTICAL AMPLIFIERS WITH MEMORIES

Specialized oscilloscopes may include digital memories in the vertical-amplifier channel. Memories provide data-domain displays, such as the example given in Fig. 1-1. The memory function is supplemented by a character-generator function (Fig. 1-12). Read-and-write random-access memories (or RAMs) are employed, as shown in Fig. 5-16. A *binary digit* (either a 0 or a 1) is called a *bit*. Four bits are termed a *nibble*. In the example of Fig. 5-16, a total of 16 memory registers is provided. (Each nibble occupies its own register in the memory. Observe that if tri-state buffers are used in the data output lines, data will not be outputted unless the tri-state buffers are driven "logic high.") Each register consists of four memory elements or cells. Each cell can remember (store) one bit—either a 0 or a 1. Each cell is accessed for *writing in* or for *reading out* by pulsing the four address lines and one of four data input lines. The address lines are decoded, so that a total of four 0–1 address combinations can access one of 16 memory registers. Digital data inputs are admitted at a desired time via AND gates when their write-enable input terminals are pulsed. Readout occurs through the data output buffer amplifiers whenever a register is accessed by the address lines.

Digital memories in the vertical-amplifier channel are also used to store digitized waveforms, and are used to reconstruct them for

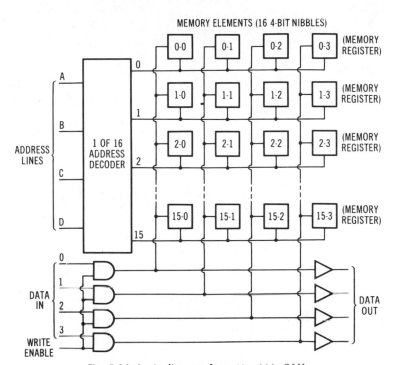

MEMORY ELEMENTS (16 4-BIT NIBBLES)

Fig. 5-16. Logic diagram for a 16-nibble RAM.

"replay" on the crt screen, as diagrammed in Fig. 5-17. A digitizer is a digital-logic device which converts analog data, such as a sine-wave waveform, into binary numbers that denote discrete amplitude levels. These discrete levels are apparent in Fig. 5-17. The binary numbers are stored in a read-and-write memory, for future use. When it is desired to "replay" the waveform, the RAM is read out automatically, and the stored binary numbers generate pulses with amplitudes corresponding to the numbers. These pulses then reconstruct the stored waveform, as shown by the example given in Fig. 5-17. Note that if the pulse train is passed through a suitable integrating circuit before it is applied to the crt, the wave envelope is recovered and the pulses per se do not appear on the screen.

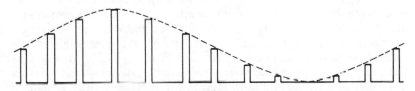

Fig. 5-17. Reconstruction of a sine wave from pulses stored in a digital memory.

CHAPTER 6

Oscilloscope Probes

TYPES OF PROBES

Various kinds of probes are required for particular applications of the oscilloscope. The more basic types of probes are as follows:
1. Direct probe—This includes open test leads, exposed cable, and coaxial cable.
2. Isolating (resistive) probe—This consists of a resistor and a coaxial cable.
3. Low-capacitance probe—This probe provides reduced input capacitance with an incidental attenuation of signal voltage.
4. Demodulator probe—This type of probe permits various tests in high-frequency circuitry.
5. High-voltage capacitance-divider probe—This type of probe attenuates high-voltage ac waveforms so that the input-voltage rating of the scope is not exceeded.
6. Special-purpose probes—These probes are used for displaying current waveforms, and for various forms of signal processing, prior to the display of the waveform.

WHY PROBES ARE NEEDED

Different kinds of oscilloscope probes are required for specific types of tests and measurements. However, it should not be supposed that probes are *always* required. For example, consider the power-supply ripple check depicted in Fig. 6-1. The output of a power supply represents a *low-impedance circuit*. In turn, the input impedance of the oscilloscope cannot load the circuit significantly. This means that we can use a direct probe made of a coaxial cable, or even a pair of open test leads to feed the ripple voltage to the vertical-input terminals of the scope. Receiver service data may specify the maximum tolerable

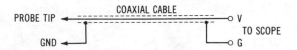

(A) Diagram of coaxial cable probe.

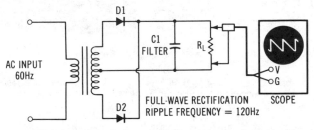

NOTE: This waveform check may also be made using open test leads.

(B) Diagram of circuit being checked.

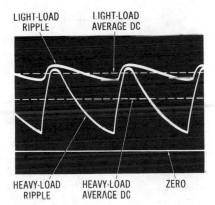

(C) Types of waveforms displayed on scope screen.

Fig. 6-1. A direct probe may be used to check power-supply ripple.

peak-to-peak ripple voltage. The reason that open test leads may be utilized to check ripple voltage is that the low impedance of the power-supply output circuit makes the test leads immune to the pickup of stray fields, which would otherwise interfere with the displayed pattern.

By way of an example, suppose that we have a pair of open test leads connected to the vertical input terminals of a scope, and that we leave the test leads lying on the bench. In turn, a 60-Hz pattern will be displayed on the scope screen, unless the vertical-gain controls are set to a low value (see Fig. 6-2). In other words, a "floating" vertical input

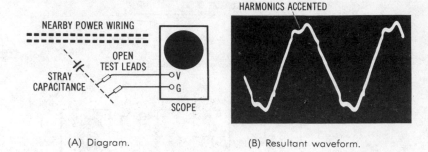

HARMONICS ACCENTED

(A) Diagram.

(B) Resultant waveform.

Fig. 6-2. Example of stray-field voltage being picked up by open test leads that are connected to the vertical input terminal of a scope.

lead to a scope will pick up stray fields, because it is capacitively coupled to the power-line wiring in the wall. Although the capacitance between a test lead and a power wire (several feet away) is very small, it is significant in this situation because the vertical input impedance of the scope is very high. This small stray capacitance has a decreasing reactance at higher frequencies, and tends to act as a high-pass filter. In other words, the higher-frequency power-line harmonics appear more prominent in the display than if the vertical input test lead were directly connected to the power line.

MANAGEMENT OF VERTICAL INPUT CAPACITANCE

An oscilloscope has 20 or 30 pF of capacitance at its vertical input terminal on the front panel, as indicated by the diagram of Fig. 6-3A. A pair of open test leads will have a stray capacitance from 5 to 50 pF, depending upon the separation of the leads. A coaxial input cable, however, will have an input capacitance of from 50 to 80 pf, depending upon its length and the type of cable. Cable capacitance adds to vertical input capacitance of an oscilloscope to form the total input capacitance. It has been noted that unshielded test leads, that are

(A) Capacitance present at the vertical input terminal on the front panel.

(B) Total input capacitance present due to use of a coaxial cable probe.

Fig. 6-3. Equivalent RC circuits present at the vertical input of a scope.

open-circuited, will pick up stray 60-Hz fields. Thus, unshielded test leads are often unsuitable for testing in high-resistance tv circuitry, or in high-resistance (high-impedance) audio circuitry. For example, FET circuits often have high internal impedance, with the result that unshielded vertical input leads to the scope will pick up excessive stray-field interference. Thus, in tv receiver circuitry, exposed test leads often pick up strong flyback pulse interference. It is standard practice, therefore, to make tv and audio waveform tests using a coaxial input cable connected to the vertical input connector of the scope.

It follows from Fig. 6-3 that the total input capacitance at the vertical input terminal of a scope, when using a coaxial cable, will be in the range from 75 to 100 pF. This arrangement is termed the direct-cable input. A direct cable can be used to check the waveform across an emitter resistor without causing any appreciable waveform distortion. On the other hand, a direct cable can impose an excessive capacitive loading at the gate or drain of an FET. An example of substantial capacitive loading on a waveform display is illustrated in Fig. 6-4. Therefore, it is general practice to use a lo-C probe with an oscilloscope. Note that a lo-C probe (when properly adjusted) does not change the low-frequency response or the high-frequency response of the scope. The probe merely reduces the effective sensitivity of the

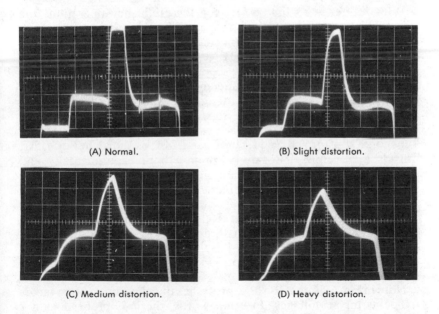

(A) Normal. (B) Slight distortion.

(C) Medium distortion. (D) Heavy distortion.

Fig. 6-4. Progressive "feathering" of the horizontal sync pulse.

scope by a factor of 10. Some lab-type lo-C probes have a 100-to-1 attenuation factor.

CONSTRUCTION AND ADJUSTMENT OF A LOW-CAPACITANCE PROBE

A general-purpose equivalent-type of vertical input circuit for an oscilloscope was shown in Fig. 6-3A. The circuit comprises 1 megohm of resistance that is shunted by 30 pF of capacitance. Next, when a coaxial input cable is connected to the scope, its equivalent general-purpose input circuit becomes 1 megohm of resistance that is shunted by 100 pF of capacitance, as illustrated in Fig. 6-3B. Now, consider how this equivalent circuit "looks" to a low-frequency voltage waveform. Since the reactance of the shunt capacitance is extremely high at very low frequencies, the vertical input circuit will "look" resistive. Next, consider how the vertical input equivalent circuit "looks" to a high-frequency voltage waveform. Inasmuch as the reactance of the shunt capacitance will be quite low at very high frequencies, the vertical input circuit will "look" capacitive. With these vertical input parameters in mind, we will consider how an RC probe can be devised that will reduce the effective input capacitance of the scope by a factor of 10, and without incurring any waveform distortion.

With reference to Fig. 6-5A, for low-frequency operation, the effective input resistance of the scope will be increased by a factor of 10, if a 9-megohm resistor is connected in series with the vertical input lead. Next, looking at Fig. 6-5B for high-frequency operation, the effective input capacitance of the scope will be decreased by a factor of 10, if an 11.11-pF capacitor is connected in series with the vertical input lead. In turn, it is reasonable to suppose that at any frequency of operation, the effective input impedance of the scope would be increased by a factor of 10, if a low-capacitance probe employed a 9-megohm resistor and an 11.11-pF capacitor, as shown in Fig. 6-5C. This is a principle that can be proven both experimentally and mathematically. The arrangement in Fig. 6-5A is called resistive voltage division, and the arrangement in Fig. 6-5B is termed capacitive voltage division.

Observe that the time constant (RC product) of the low-capacitance probe in Fig. 6-5C is 99.99×10^{-6}, and that the time constant of the scope input circuit is 100×10^{-6}. It is this equality of time constants for both probe and scope that provides a distortionless display of waveforms.

Note in Fig. 6-5C that the lo-C probe has the same attenuation factor for dc as for ac voltages. In other words, the lo-C probe does not disturb the normal response of a dc scope. However, a word of

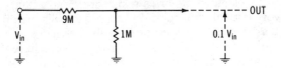

(A) At low frequencies, a 9-megohm resistor divides the input signal voltage $\frac{1}{10}$.

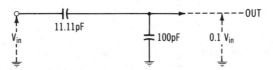

(B) At high frequencies, an 11.11-pF capacitor divides the input signal voltage $\frac{1}{10}$.

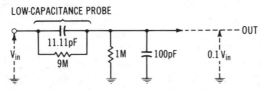

(C) At any frequency, a 9-megohm resistor shunted by an 11.11-pF capacitor divides the input voltage $\frac{1}{10}$.

Fig. 6-5. Principles of a low-capacitance probe circuit.

caution. Not all oscilloscopes have a 1-megohm input resistance and a 30-pF (or 100 pF with coaxial cable) input capacitance. Therefore, the R and C values in a lo-C probe must be correctly selected for the oscilloscope with which the probe will be used. Fig. 6-6 illustrates a combination direct and low-capacitance probe. Virtually all lo-C probes provide an adjustment of series capacitance, although few

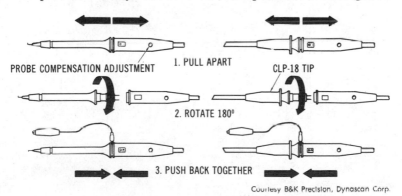

Fig. 6-6. A combination direct and low-capacitance probe.

provide any adjustment of series resistance. Capacitance values are comparatively critical with regard to waveform distortion. Note also that there is a practical consideration involved in a design choice of the 1/10 attenuation factor—if a direct probe is substituted for a lo-C probe, the operator simply moves the decimal point in his calibration value one place to the left (see Fig. 6-7). This is a simple process, compared with an arbitrary arithmetical calculation.

The adjustment of the trimmer in a lo-C probe can be checked as shown in Fig. 6-8. A 10–kHz sine-wave voltage from a generator is applied to a scope via a direct probe, and then via a lo-C probe. The first check is made using the "10" step of the vertical attenuator, and the second check is made using the "1" step of the vertical attenuator. If the lo-C probe is in correct adjustment, both displays will show an equal height on the screen. On the other hand, if a disparity is observed in displays from the two probes, the trimmer in the lo-C probe should be adjusted as required. Note that when the test frequency is reduced to 100 Hz, the response of the lo-C probe should be the same as at a 10-kHz test frequency. In case that the height of the 100-Hz pattern is different from the height of the 10-kHz pattern, it indicates that the resistor in the lo-C probe is off value. Also, a word of caution—a good

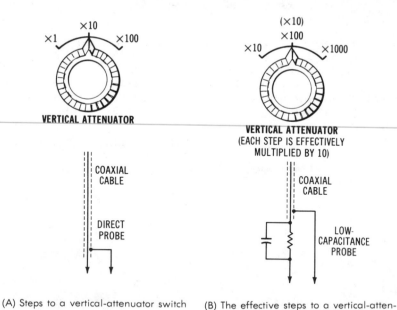

(A) Steps to a vertical-attenuator switch when using a direct probe.

(B) The effective steps to a vertical-attenuator switch when using a low-capacitance probe.

Fig. 6-7. A low-capacitance probe reduces the vertical gain by a factor of 10.

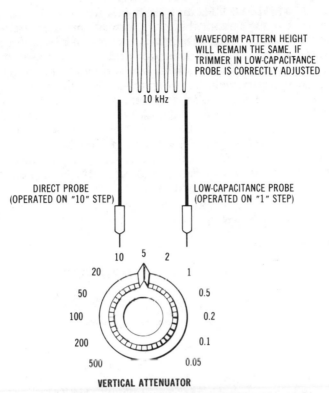

Fig. 6-8. Checking the adjustment of trimmer in a lo-C probe.

Courtesy B&K Precision, Dynascan Corp.

audio oscillator provides the same output voltage at 100 Hz as at 10 kHz. However, an economy-type oscillator may lack uniform output; this possibility can be checked using the direct probe.

Construction of low-capacitance probes for use with wide-band scopes is more critical than those used with narrow-band scopes. For example, a low-capacitance probe that performs satisfactorily with a 4-MHz scope may produce excessive overshoot and ringing on fast-rise waveforms when used with a 15-MHz scope. Therefore, low-capacitance probes can lead to unexpected difficulties if designed for use with comparatively low-performance oscilloscopes.

CONSTRUCTION AND OPERATION OF DEMODULATOR PROBES

Demodulator systems may employ either series detectors or shunt detectors. Both types find application in demodulator probes. Troubleshooting with the oscilloscope may involve tests in circuits operating at 20 MHz, 40 MHz, or even higher frequencies. However, service-type scopes seldom have a vertical-amplifier frequency response beyond 5 MHz. Therefore, to display waveforms in high-frequency circuitry, a demodulator probe must be used (a lo-C probe has no response in this situation). A demodulator probe is a simple detector arrangement that operates on the same basic principle as the picture detector in a tv receiver; a diode rectifier and its associated circuit recover the modulation envelope from a high-frequency amplitude-modulated carrier. Note that the modulation envelope contains low frequencies which are within the response range of the vertical amplifier of the scope. Basic series- and shunt-detector configurations are shown in Fig. 6-9; the shunt arrangement has a comparatively high output impedance.

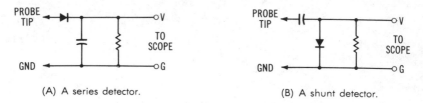

(A) A series detector. (B) A shunt detector.

Fig. 6-9. Basic detector arrangements.

Demodulation action of a series detector is shown step-by-step in Fig. 6-10. Observe that R1 and C2 operate as a partial filter for the rectified signal from the detector. In other words, the time constant of R1C2 is chosen sufficiently long so that individual half-cycles of the rf carrier cannot be "followed." In turn, the charge on C2 effectively "follows" the modulation envelope of the input waveform. This voltage output waveform is not entirely smooth, but tends to have a residual sawtooth outline. However, the modulation envelope of the input waveform is recovered for all practical purposes. Note particularly that when the R1C2 time constant is too long, the voltage output waveform does not "follow" the rise and fall of the carrier as it should. Instead, the valleys in the modulation envelope tend to be "left behind." It is this factor which determines the highest modulating frequency to which a detector can satisfactorily respond.

Simple Demodulator Probe

The simplest demodulator probe design is shown in Fig. 6-11. It consists of a semiconductor diode·connected in series with a coaxial cable. As depicted in Fig. 6-11A, the cable capacitance serves as a charging capacitor for the diode—it operates as a low-pass filter. The dc charge that builds up on the capacitor proceeds to discharge through the 1-megohm resistance of the vertical attenuator and also through

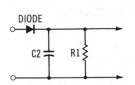

(A) Circuit configuration.

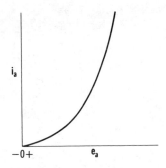

(B) $i_a - v_a$ characteristic for a crystal diode.

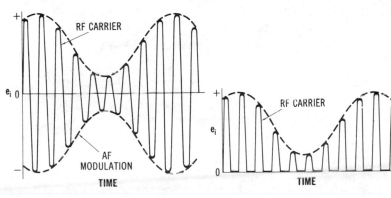

(C) Voltage input waveform.

(D) Rectified signal-diode current with capacitor.

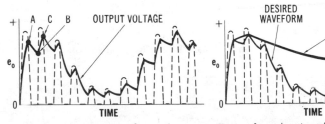

(E) Voltage output waveform when C2 is added.

(F) Waveform showing effect of too large a time constant.

Fig. 6-10. Step-by-step demodulation action of a series detector.

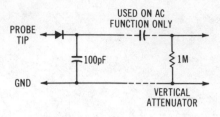

(A) Equivalent circuit.

(B) Probe with a negative-going output.

(C) Probe with a positive-going output.

Fig. 6-11. Simplest type of demodulator probe design.

the back (reverse) resistance of the diode. This probe has limited usefulness at very high frequencies, because the coaxial cable does not "look" like a simple capacitor at higher frequencies. The cable develops standing waves. In turn, it has a very high input impedance at one resonant frequency, and has a very low input impedance at another of its resonant frequencies. Therefore, a more elaborate probe circuitry is preferred in practical troubleshooting procedures. Note that if the diode is polarized one way, a negative-going output is obtained, and if the diode is polarized the other way, a positive-going output is obtained. This is shown in Fig. 6-12.

Shunt-Detector Probe

A standard demodulator probe circuit is exemplified in Fig. 6-13.

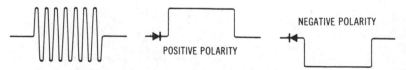

Fig. 6-12. Demodulated output waveform is "turned upside down" if polarity of detector diode is reversed.

This is a shunt-detector configuration with a two-section RC filter; the coaxial cable provides the capacitance for the second section. Note that the 220K resistor serves as a filter component, and also isolates the coaxial cable from the high-frequency circuit, thereby avoiding development of standing waves. Its response characteristics are:

Carrier frequency range—500 kHz to 200 MHz.

Output (envelope) frequency range—30 Hz to 5 kHz.

Probe input capacitance (approximate)—2.25 pF.

Probe input resistance (approximate) is:

    25 kilohms at 500 kHz,
    23 kilohms at 1 MHz,
    21 kilohms at 5 MHz,
    18 kilohms at 10 MHz,
    10 kilohms at 50 MHz,
     5 kilohms at 100 MHz,
    4.5 kilohms at 150 MHz,
    2.5 kilohms at 200 MHz.

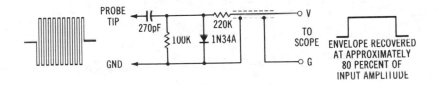

Fig. 6-13. A standard demodulator probe circuit.

Maximum ac voltage input—20 rms volts, 28 peak volts. Since the envelope frequency capability of the probe extends only to 5 kHz, a horizontal sync pulse cannot be reproduced, although a vertical sync pulse is reproduced when the demodulator probe is applied in a tv i-f circuit. However, a vertical sync pulse is distorted to the extent that its serrations and equalizing pulses are "wiped out." Thus, the reproduced vertical sync pulse has a superficial similarity to a horizontal sync pulse in this situation (see Fig. 6-14). The demodulator probe has an input capacitance of approximately 2.25 pF and an input resistance of about 10 kilohms at 50 MHz. Accordingly, tv i-f circuits are substantially loaded and somewhat detuned by a probe application. In turn, the demodulator probe serves as a signal-tracing device, but it is not a reliable indicator of the signal-voltage level. However, a demodulator probe is an accurate *relative signal-level indicator* when used in low-impedance circuits. For example, a demodulator probe is a reliable indicator to check the uniformity of an output from a sweep generator (Fig. 6-15).

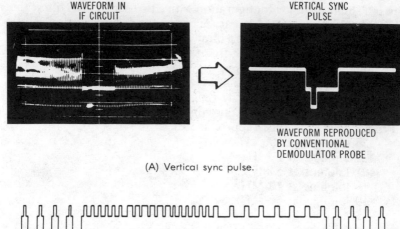

WAVEFORM IN
IF CIRCUIT

VERTICAL SYNC
PULSE

WAVEFORM REPRODUCED
BY CONVENTIONAL
DEMODULATOR PROBE

(A) Vertical sync pulse.

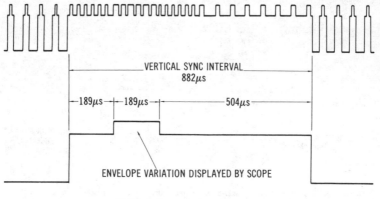

VERTICAL SYNC INTERVAL
882μs

―189μs―|―189μs―|――――504μs――――

ENVELOPE VARIATION DISPLAYED BY SCOPE

(B) Waveform timing diagram.

Fig. 6-14. Only the lower video frequencies are reproduced by a conventional demodulator probe.

Medium-Impedance Demodulator Probe

A medium-impedance demodulator probe, depicted in Fig. 6-16, is sometimes used to avoid substantial waveform distortion. For example, the demodulator probe shown in Fig. 6-13 reproduces vertical sync pulses in the video signal, but this probe practically "wipes out" the horizontal sync pulses. On the other hand, the demodulator probe shown in Fig. 6-16 reproduces both the vertical sync pulses and a reasonable replica of the horizontal sync pulses in the video signal. However, a medium-impedance demodulator probe imposes more circuit loading and it attenuates the signal to a greater extent than a higher-impedance probe, such as the one depicted in Fig. 6-13. Thus, the probe shown in Fig. 6-16 is a compromise design between circuit loading and waveform reproduction. Both types of

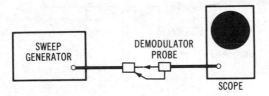

(A) Equipment setup.

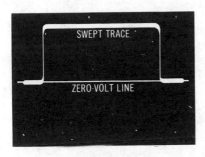

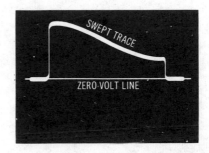

(B) An ideal screen pattern.

(C) A deficient pattern.

Fig. 6-15. Demodulator probe is used to check the uniformity of an output from a sweep generator.

demodulator probes are sometimes called *traveling detectors*, because they can be used to trace a signal stage-by-stage through an i-f amplifier section. As a practical note, signal tracing at the input (and occasionally at the output) of the first i-f stage may be impossible unless a scope with high sensitivity, such as 10 mV/in, is used.

RESISTIVE "ISOLATING" PROBE

A resistive "isolating" probe is a simple arrangement that consists of a resistor connected in series with the coaxial input cable to the scope, as shown in Fig. 6-17A. Although called an "isolating probe," it is technically a low-pass RC filter section. It can be compared with an integrating circuit. This type of probe is used only in sweep-alignment procedures. Low-pass filtering action serves to sharpen the beat-marker (pip) indication on a response curve. It also serves to minimize noise (fuzz) on the curve when checking the response of low-level circuits. In most situations, a 50K resistor is suitable. However, if the resistor is too large, the marker indication on the side of a response curve will be displaced (due to time delay). On the other hand, if the resistor is too small, the beat-marker indication will be broader than necessary (high beat frequencies are passed.)

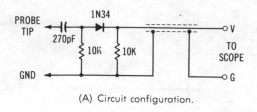

(A) Circuit configuration.

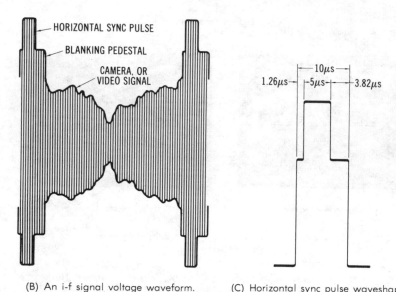

(B) An i-f signal voltage waveform.　　(C) Horizontal sync pulse waveshape.

Fig. 6-16. A medium-impedance demodulator probe that can reproduce horizontal sync pulses.

DOUBLED-ENDED DEMODULATOR PROBE

A double-ended (push-pull) demodulator probe, as depicted in Fig. 6-18, is preferred for checking the termination of a transmission line. (If a coaxial-cable termination is to be checked, a conventional single-ended demodulator probe is utilized.) This sweep test is based on the development of standing waves that occur in case the line is incorrectly terminated. Standing waves result in a change of output voltage over the swept band. On the other hand, if the line is correctly terminated, no standing waves occur, and the scope pattern is the same as if the demodulator probe were applied at the output terminals of the sweep generator. The same method can be used to check whether a transmission line is correctly matched by an antenna, and whether *matching stubs* and similar devices are in proper adjustment. If the line is terminated by an antenna, the double-end demodulator probe may be connected at the generator end of the line.

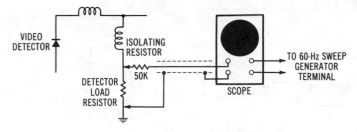

(A) Connections made to circuit for testing.

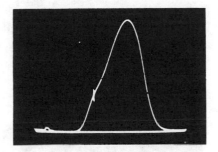

(B) An i-f response curve with beat marker.

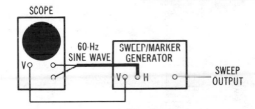

(C) A 60-Hz sine wave will deflect the scope pattern.

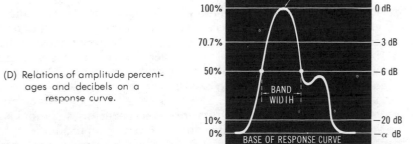

(D) Relations of amplitude percentages and decibels on a response curve.

NOTE: Bandwith is measured between the −6 dB points.

Fig. 6-17. Application of a resistive "isolating" probe in sweep-alignment procedures.

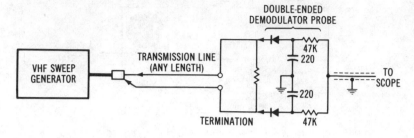

(A) Circuit configuration.

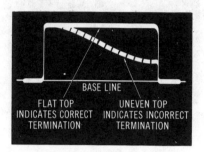

(B) Use of scope pattern to check lead-in termination.

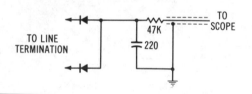

(C) An alternate probe circuit.

Fig. 6-18. A double-ended demodulator probe.

HIGH-VOLTAGE CAPACITANCE-DIVIDER PROBE

Although used primarily in laboratories, high-voltage capacitance-divider probes also find an occasional application in tv troubleshooting procedures. High peak-to-peak voltages are encountered in the horizontal-sweep section of a tv receiver. These voltages will arc through a lo-C probe, damaging both probe and scope. A special probe, therefore, is required to test these high ac voltages. A typical circuit is shown in Fig. 6-19. This is a capacitance-divider circuit arrangement. When two capacitors are connected in series, an applied ac voltage drops across the capacitors in inverse proportion to their capacitance values. Thus, if one capacitor has 99 times the capacitance of the other, 0.01 of the applied voltage is dropped across the larger capacitor. In turn, the smaller capacitor requires a high-voltage rating.

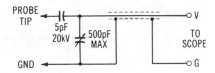

Fig. 6-19. Typical high-voltage capacitor-divider probe.

The standard attenuation factor of a high-voltage capacitor-divider probe is 100 to 1, and is set by a trimmer capacitor. This 100-to-1 attenuation factor is used to tie the probe attenuation in with the attenuation of the step attenuator of the scope. The probe attenuates horizontal sweep-circuit waveforms to 0.01 of their source-voltage value, thus protecting the scope against damage. Fig. 6-20 shows a probe circuit where a spark gap provides overload protection.

The high-voltage capacitance-divider probe is uncompensated, and is, therefore, useful only at horizontal-deflection frequencies. Thus, vertical-frequency waveforms would be distorted. However, this is not a drawback, because the attenuation factor of the probe restricts its application to the horizontal-sweep circuitry. The reason that the 100-to-1 probe is unsuitable for vertical-section tests is seen in Fig. 6-21. Observe that the probe capacitors do not stand alone, but work

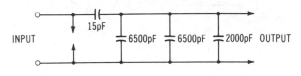

(A) Schematic diagram.

(B) Physical appearance.

Fig. 6-20. Another capacitor-divider probe.

111

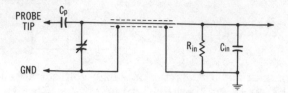

(A) Configuration when connected to vertical input of scope.

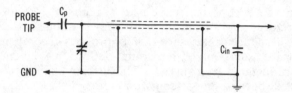

(B) Equivalent circuit at high frequencies.

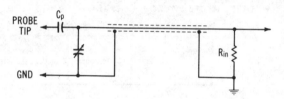

(C) Equivalent circuit at low frequencies.

Fig. 6-21. High-voltage capacitance-divider probe and its load circuit.

into the vertical-input impedance (R_{in} and C_{in}) of the scope. The shunt resistance R_{in} can be neglected at horizontal-deflection frequencies, because this resistance value is very high compared with the low reactance of the input capacitance. On the other hand, at vertical-deflection frequencies, the shunt resistance R_{in} has a value on the same order as the reactance of the input capacitance. In turn, the capacitance-divider probe now acts as a differentiator of the vertical-frequency waveform, and the display is badly distorted.

CLAMP-AROUND CURRENT PROBE

Clamp-around current probes, such as pictured in Fig. 6-22, are used chiefly in laboratory work. Of course, a current probe can be used in electronic servicing, if desired. As an illustration, if the probe is clamped around a lead to the horizontal-deflection coils in a tv

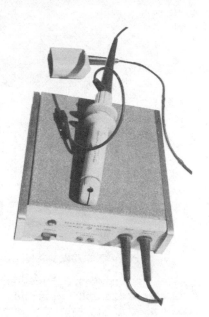

Fig. 6-22. A clamp-around current probe for displaying current waveforms.

receiver, the current sawtooth waveform will be displayed. This type of probe is called an active device. The probe itself contains a miniaturized current transformer, of which the lead being tested becomes the equivalent of a one-turn primary. Next, the output from the current transformer is stepped up by a transistor amplifier (with self-contained batteries). The output from the amplifier is applied to the vertical-input terminals of a scope. A typical current probe has a sensitivity of 1 mV per mA, and a frequency range from 60 Hz to 4 MHz.

CHAPTER 7

Supplementary Equipment

TYPES OF SUPPLEMENTARY OSCILLOSCOPE EQUIPMENT

Many types of supplementary equipment are used with oscilloscopes. For example, semiconductor quick checkers and curve tracers, various signal generators, pickup devices, signal processors, transducers, and so on, are used with the oscilloscope in various kinds of tests and measurements. Thus, a microphone is used with an oscilloscope in tone-burst tests of hi-fi speakers. (A microphone is a transducer.) A swept harmonic analyzer is used with an oscilloscope in the spectrum analysis of audio waveforms. (The swept harmonic analyzer is a signal processor.) A pickup assembly (clamp-around probe) is used with the oscilloscope in ignition analysis. Many types of a-m and fm generators are utilized with the oscilloscope; specialized generators are also employed. Color-bar generators and fm stereo generators are some familiar examples of specialized generators that are used with the oscilloscope. Digital word generators (a specialized form of pulse generator) may be used with the oscilloscope in design work, and in some types of digital-equipment troubleshooting.

SEMICONDUCTOR QUICK CHECKER

A simple semiconductor quick checker circuit is shown in Fig. 7-1. It consists of a 10-to-1 stepdown transformer and four resistors. Two test leads are also provided. Observe that when the test leads are short-circuited, a vertical trace is displayed on the scope screen (Fig. 7-1B). On the other hand, when the test leads are open-circuited, a nearly horizontal line is displayed on the scope screen. (The 10-to-1 stepdown transformer, T1, supplies a secondary test voltage of 11.7 volts rms on open circuit.) If the test leads are connected across various values of resistance, diagonal lines with various slopes will be

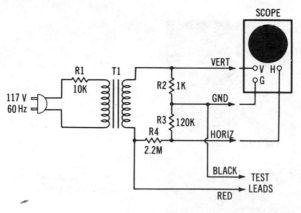

(A) Circuit configuration.

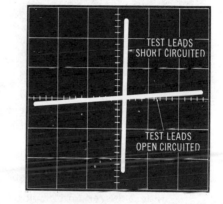

(B) Displays resulting from open-circuited and short-circuited test leads.

Fig. 7-1. A simple semiconductor quick check setup.

displayed, as exemplified in Fig. 7-2. However, if the test leads are connected across a germanium diode or a silicon diode, waveforms such as those depicted in Figs. 7-3A and 7-3B will normally be displayed. Zener diodes will normally produce waveforms such as those shown in Figs. 7-3C through 7-3E. The polarity identification for different types of diodes is given in Fig. 7-4. Small-signal characteristic curves for Ge and Si diodes are shown in Fig. 7-5. Defective diodes produce straight or horizontal lines (or diagonal lines as shown in Fig. 7-2) on the scope screen.

TESTING TRANSISTORS

A bipolar transistor is tested as if it were two diodes connected in series with opposing polarities. Thus, if the transistor is a pnp type, the

115

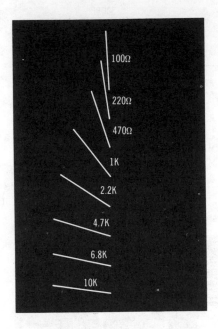

Fig. 7-2. Slopes of screen traces for corresponding values of resistances.

base material is a cathode, whereas, the emitter and collector are anodes. Some waveforms normally produced by transistors are depicted in Fig. 7-6. If a transistor is defective, a straight or horizontal line will be displayed (or else, diagonal lines) as was shown in Fig. 7-2. The terminals of most transistors and their polarity type (pnp or npn) are checked as follows:

1. Connect the red and black test leads at random to the transistor terminals. When a waveform is obtained as shown in Figs. 7-6A and 7-6B, the leads will be connected across the base-emitter junction of the transistor.
2. Next, connect the red test lead to the collector terminal.
3. Then, connect the black lead alternately to the two remaining terminals of the transistor. A greater vertical deflection will be obtained when the black lead is connected to the base of the transistor.
4. With the black lead connected to the base, vertical deflection will be downward at the right and upward at the left for most npn transistors.

Note that *most transistors show a zener-type conduction* through their base-emitter junction. However, for those transistors that do not have zener conduction, such as germanium types, the foregoing first step cannot be carried out, and the following procedure should be followed instead.

1. Connect the red and black test leads at random to the transistor terminals. When an open circuit (horizontal line) is displayed on the screen, the test leads are connected to the collector and emitter (a small conduction may be observed near the voltage crossover point).
2. Now, connect the black test lead to the base of the transistor.
3. With the black test lead connected to the base, a pnp transistor will produce a downward deflection at the right, and an npn transistor will produce an upward deflection at the left of the pattern.

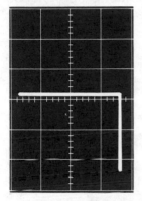

(A) Black lead connected to cathode of diode.

(B) Black lead connected to anode of diode.

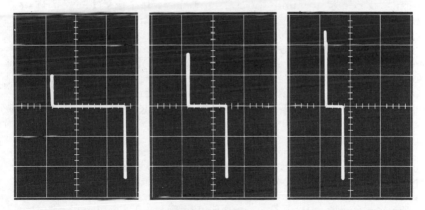

(C) Waveform of a 30-volt zener diode.

(D) Waveform of a 15-volt zener diode.

(E) Waveform of a 7-volt zener diode.

Fig. 7-3. Normal waveforms that are displayed when the test leads are connected across a diode.

117

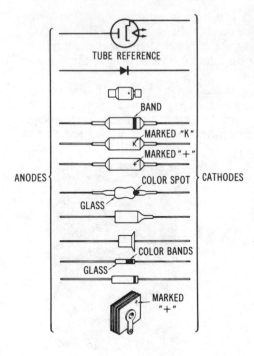

TUBE REFERENCE

BAND

MARKED "K"

MARKED "+"

ANODES

COLOR SPOT

CATHODES

GLASS

COLOR BANDS

GLASS

MARKED "+"

Fig. 7-4. The schematic representation and polarity markings of tube, germanium, silicon, and selenium diodes.

SEMICONDUCTOR CURVE TRACER

A semiconductor curve tracer, such as depicted in Fig. 7-7A, will display a complete family of collector characteristics (Fig. 7-7B) or base characteristics (Fig. 7-7C) on the scope screen. The horizontal-deflection voltage is a succession of 60-Hz half-sine waves which vary the collector potential of the transistor under test from zero to some peak value, and then back to zero. The peak voltage is chosen by the operator. At the end of each half-sine excursion, the base-bias voltage is increased by a step generator; the value of step voltage is chosen by the operator. In turn, the scope displays the collector family of characteristics. The base family of characteristics can also be displayed (Fig. 7-7C).

Fig. 7-8 shows how the pattern on the scope screen is developed by a curve tracer. On the first step of the staircase waveform, a horizontal trace occurs; this is followed by a higher horizontal trace on the second step, and by a still higher horizontal trace on the third step. A six-step collector-family pattern is exemplified in Fig. 7-9. Displayed patterns can be evaluated by checking against the data sheets or manuals of the semiconductor manufacturers. The beta value of a transistor can be determined from its screen pattern as shown in Fig. 7-10. Beta is equal to the output current change divided by the input current change.

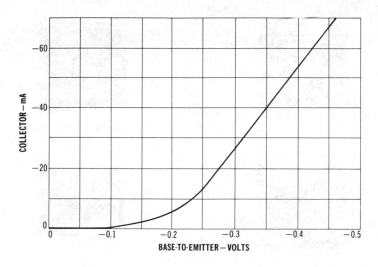

(A) Germanium diode.

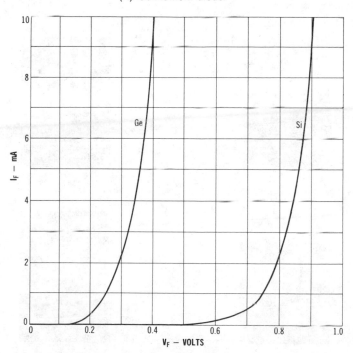

(B) Comparative curves for silicon and germanium diodes.

Fig. 7-5. Normal small-signal curves for semiconductor diodes.

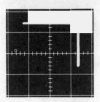

(A) Pnp base-emitter waveforms.

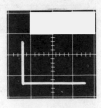

(B) Npn base-emitter waveforms.

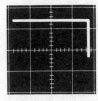

(C) Pnp collector-base waveform with red lead connected to collector.

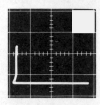

(D) Npn collector-base waveform with red lead connected to collector.

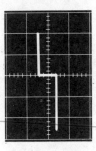

(E) Black lead is connected to the base.

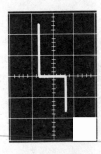

(F) Black lead is connected to the base.

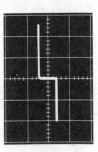

(G) Black lead is connected to the emitter.

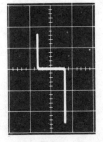

(H) Black lead is connected to the emitter.

Fig. 7-6. Waveforms normally produced by transistors.

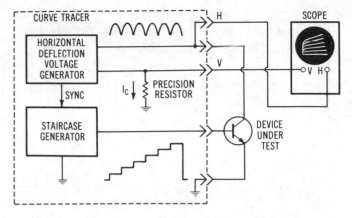

(A) Test setup.

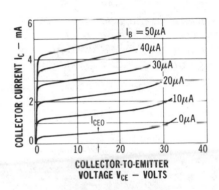

(B) Collector characteristic curves.

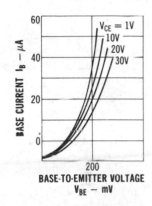

(C) Base characteristic curves.

Fig. 7-7. A semiconductor curve tracer.

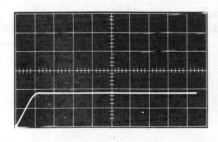

(A) Horizontal trace on first step of staircase.

(B) Horizontal traces on next two higher steps.

Fig. 7-8. Development of a screen pattern by a curve tracer.

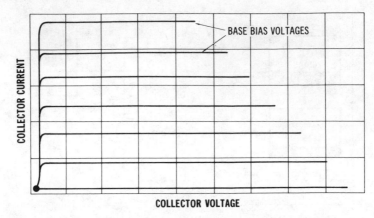

Fig. 7-9. A six-step display of collector characteristic curves.

Thus, in the example given in Fig. 7-10, the transistor beta value is 180. Observe that the output impedance of a transistor can also be determined from its screen pattern. This is shown in Fig. 7-11. The output impedance is equal to the output voltage change divided by the output current change. Thus, in the example of Fig. 7-11, the output impedance value is 31,000 ohms.

OSCILLOSCOPE TESTS USING AUDIO GENERATORS

Many oscilloscope tests are made in combination with audio generators. For example, a frequency response test of an audio

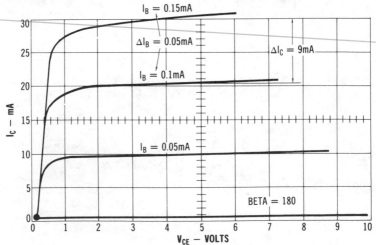

Fig. 7-10. Determination of the beta value of a transistor.

122

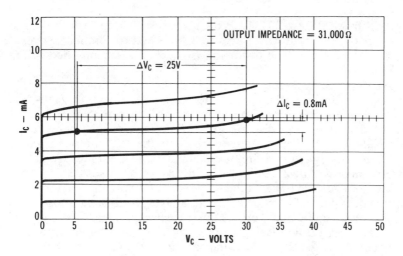

Fig. 7-11. Determination of the output impedance of a transistor.

amplifier is made as shown in Fig. 7-12. Note that this test also provides a measure of the amplifier gain. In other words, the ratio of the output voltage to the input voltage is equal to the voltage gain of the amplifier. Note that the amplifier is terminated in its rated value of load resistance by a power-type resistor with an adequate wattage capability. (The resistor will get quite hot when a large audio amplifier is under test.) The operator may also inspect the oscilloscope pattern for any evidence of clipping or crossover distortion. A high-fidelity amplifier normally will have a frequency response that is flat within

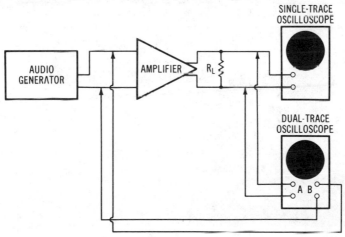

Fig. 7-12. Test setup for checking amplifier frequency response.

± 1 dB from 20 Hz to 20 kHz. A dual-trace oscilloscope will also permit the input and output waveforms to be checked for proper layover (a lack of distortion). The distortion is checked as exemplified in Fig. 7-13.

An fm stereo generator is a specialized type of audio generator. It checks an fm decoder (of a stereo fm receiver) for left- and right-channel separation. This test is made as shown in Fig. 7-14. This is called a vectorscope display. The "L" output signal is applied to the vertical-input channel of the scope, and the "R" output signal is applied to the horizontal-input channel of the scope. In turn, a pattern is obtained as shown in Fig. 7-15. In case of ideal separation, there would be a vertical trace on the scope screen. In practice, a normal fm stereo decoder will provide approximately a 30-dB separation; this is indicated by a diagonal vertical line that has an angle of slightly less than 5° from the vertical. In the event that there is no separation between the L and R signals, a diagonal line is displayed which makes a 45° angle from the vertical. Zero separation corresponds to monophonic reproduction.

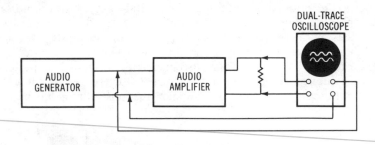

(A) Test setup.

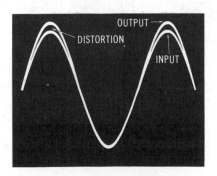

(B) Superimposed input/output patterns.

Fig. 7-13. Checking for audio distortion with a dual-trace oscilloscope.

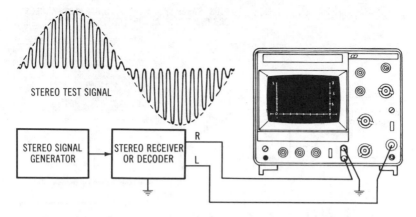

STEREO TEST SIGNAL

STEREO SIGNAL
GENERATOR

STEREO RECEIVER
OR DECODER

R

L

Fig. 7-14. Using an oscilloscope to check stereo separation.

Square-Wave Test

An audio amplifier is checked for square-wave response with an oscilloscope and a square-wave generator as shown in Fig. 7-16. Note that the amplifier should be driven to maximum-rated power output on the basis of

$$P = \frac{(E_p)^2}{R_L}$$

where,

P is the power in watts,
E_p is the peak voltage of the square wave,
R_L is the value of the load resistor.

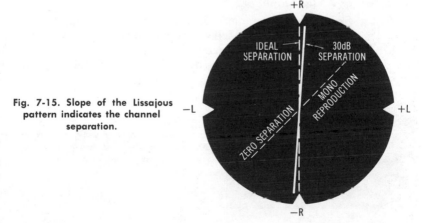

Fig. 7-15. Slope of the Lissajous pattern indicates the channel separation.

+R

IDEAL
SEPARATION

30dB
SEPARATION

MONO
REPRODUCTION

−L

+L

ZERO SEPARATION

−R

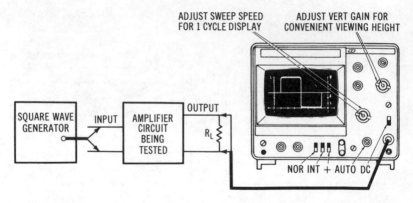

Fig. 7-16. Test setup for the square-wave testing of a hi-fi amplifier.

In other words, the peak voltage and the rms voltage of a square wave have the same value—the peak voltage is used to calculate the amplifier power output. Manufacturers generally rate hi-fi amplifiers for square-wave response at a 2-kHz repetition rate (Fig. 7-17). It is evident that the rise time and fall time are sufficiently slow so that the leading and trailing edges of the reproduced square wave are visible. The rise time of the amplifier is related to its high-frequency cutoff by the equation

$$T_r = \frac{1}{3}f_c$$

where,

T_r is the rise time,

f_c is the -3 dB cutoff frequency of the amplifier.

Pulse Test of Audio Amplifier

Most hi-fi amplifiers have music-power ratings that are higher than their rms power ratings. The rms power ratings are based on a steady-state output with respect to the rms voltage of a sine wave. On the other hand, music-power ratings denote the ability of an amplifier

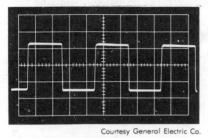

Fig. 7-17. Scope pattern showing the rated 2-kHz square-wave response for a 10-watt hi-fi amplifier.

126

to process a sudden surge (attack) waveform without distortion. With reference to Fig. 7-18, music-power ratings are determined by observing the response of an amplifier to pulse voltages. The standard

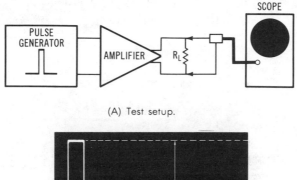

(A) Test setup.

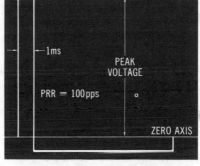

(B) Undistorted output pulse.

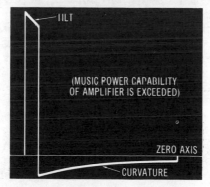

(C) Distorted output pulse.

Fig. 7-18. Pulse test of a hi-fi amplifier for determination of music-power capability.

test is made with a pulse width of 1 millisecond and with a pulse repetition rate of 100 pulses per second. The amplitude of the generator output signal is gradually increased until the scope pattern shows evidence of tilt and curvature in the reproduced pulse waveform. Then, the generator output is reduced until there is no visible indication of pulse distortion. In turn, the music-power value is given by

$$\frac{(E_p)^2}{R_L}$$

where,
E_p is the peak voltage of the pulse,
R_L is the value of the load resistor.

OSCILLOSCOPE TESTS USING HIGH-FREQUENCY GENERATORS

Various types of tests are made with oscilloscopes in combination with high-frequency generators. For example, an fm receiver is usually aligned using sweep and marker generators, with indication of its frequency response being monitored by an oscilloscope. In the example of Fig. 7-19, an i-f sweep-and-marker signal is applied at the input of the mixer stage. Note that an rf sweep-and-marker signal could be applied at the input of the rf amplifier to obtain an overall rf/if frequency-response curve. When the oscilloscope is connected at the input of the limiter, the demodulated i-f signal is displayed on the scope screen. In other words, the limiter transistor operates nonlinearly and develops the envelope of the i-f signal. When the oscilloscope is suitably connected into the fm detector circuit, the "S" curve (detector frequency response) is displayed on the scope screen.

A television receiver contains an fm receiver as a subsection. In other words, the intercarrier sound section of the tv is basically an fm receiver. However, the intercarrier i-f circuits are tuned to 4.5 MHz, whereas, the i-f circuits in Fig. 7-19 were tuned to 10.7 MHz. Alignment procedure for an intercarrier sound section is essentially the same as for an fm receiver. The intercarrier sound section sometimes develops trouble symptoms involving sync buzz. This is a harsh 60-Hz rasping noise that usually results from the intermodulation of the tv sound signal with the vertical sync component of the picture signal. An oscilloscope can be used to trace and to analyze sync buzz. The buzz waveform usually appears as shown in Fig. 7-20. Its origin in the majority of cases is in a video-amplifier stage that is operating nonlinearly.

Oscilloscopes are used in combination with keyed rainbow color-bar generators to display vectorgrams, as shown in Fig. 7-21. A vectorgram is a specialized form of Lissajous figure which shows

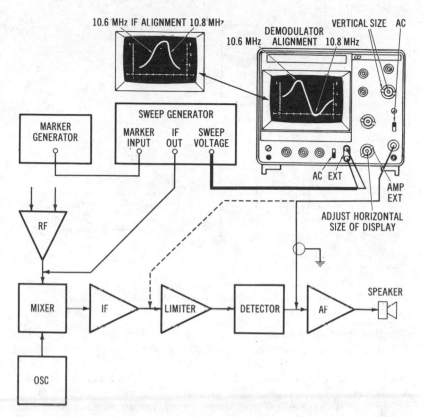

(A) Test setup.

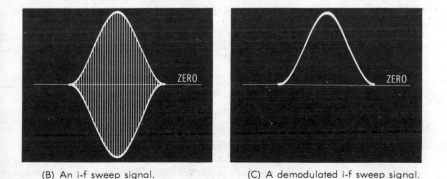

(B) An i-f sweep signal.

(C) A demodulated i-f sweep signal.

Courtesy B&K Precision, Dynascan Corp.

Fig. 7–19. Sweep alignment of a fm receiver.

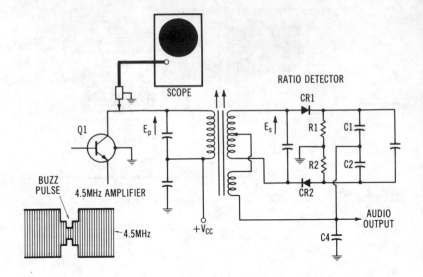

Fig. 7-20. An oscilloscope can be used to trace sync-buzz interference.

chroma-signal amplitudes at various phases. Ten "petals" appear at 30° intervals in a normal vectorgram. Vectorgrams are usually produced by chroma signals from the R-Y and B-Y channels, although the technique is not limited to this particular arrangement. In other words, R-Y/G-Y vectorgrams or B-Y/G-Y vectorgrams can be displayed, if desired. Some color-tv receivers utilize X and Z demodulation; these phases are different from the R-Y and B-Y phases. In turn, if the scope is connected at the outputs of the X and Z demodulators, the pattern will be elliptical instead of circular. However, when the scope is moved ahead to the R and G terminals of the color picture tube, the X and Z signals will normally have been matrixed into R-Y and B-Y signals, so that a circular vectorgram is then displayed.

OSCILLOSCOPE TESTS USING DIGITAL WORD GENERATORS

Oscilloscopes may be used with digital word generators to check the operation of digital equipment. A very simple example is shown in Fig. 7-22. Here, the operation of an AND gate is being tested. One input of the AND gate is energized by a continuous train of pulses, while the other input is energized by widely separated single pulses. If the gate is operating normally, an output pulse will be displayed only when both of the gate inputs are simultaneously energized by pulses. Gates are the basic building blocks of all digital logic configurations. A typical

130

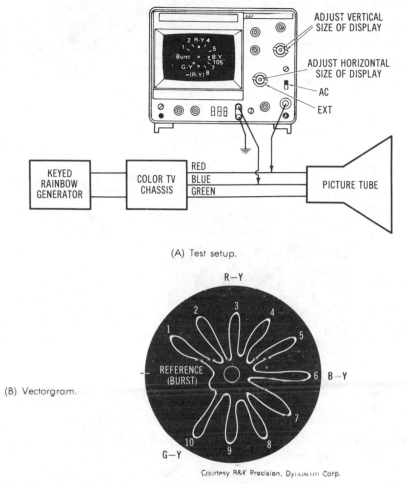

(A) Test setup.

(B) Vectorgram.

Courtesy B&K Precision, Dynascan Corp.

Fig. 7-21. An oscilloscope is used with a color-bar generator to produce vectorgrams.

adder will contain seven gates; a four-place adder employs 28 gates. One of the trouble conditions encountered in a complex digital system is spurious pulses called *glitches*. A glitch can result from various causes, one of which is progressive waveform deterioration through the system. Glitch waveforms may be very narrow and may occur intermittently. In turn, high-performance oscilloscopes are often required in order to capture a "random" narrow glitch and to retain it on screen.

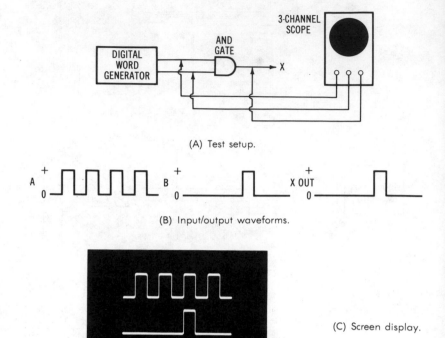

(A) Test setup.

(B) Input/output waveforms.

(C) Screen display.

Fig. 7-22. Check of an AND gate operation using a 3-channel scope.

Adjusting and Servicing the Oscilloscope

MAINTENANCE AND SERVICING REQUIREMENTS

All oscilloscopes eventually require maintenance or servicing. Devices and components gradually deteriorate, and oscilloscope performance drifts "out of spec." Sometimes catastrophic failure occurs; the oscilloscope may not respond to an input signal, the displayed pattern may be distorted, or the screen may be dark. In such a case, servicing procedures are required. There is no sharp dividing line between maintenance and servicing operations. For example, if a checkup shows that the vertical step attenuator is out of calibration, a readjustment of the vertical-gain maintenance control (on the scope chassis) may be all that is required. On the other hand, if the vertical-gain maintenance control is out of range, servicing procedures are then in order. It is good practice to periodically check the performance of an oscilloscope to determine whether it may be "out of spec." Most maintenance checkups can be accomplished with ordinary instruments and the equipment available in tv service shops.

ROUTINE MAINTENANCE CHECKS

A typical triggered-sweep oscilloscope is rated for an input impedance of 1 megohm shunted by 30 pF of capacitance. With reference to Fig. 5-13, the input impedance of an oscilloscope is determined by the characteristics of the vertical step attenuator. The value of input impedance is normally the same on each step of the attenuator. Incorrect values of input impedance are ordinarily caused by capacitor leakage, by defective resistors, or by deteriorated switches. To measure the vertical-input resistance, a 60-Hz voltage may be applied to the vertical-input terminal of the scope, as shown in Fig. 8-1. The amount of vertical deflection is first noted with no resistance in the circuit. Next, a potentiometer is used to insert sufficient series resistance so that the amount of vertical deflection is reduced to one-half. Then, the potentiometer is disconnected and its resistance value is measured with an accurate ohmmeter. This value is equal to the vertical-input resistance of the oscilloscope.

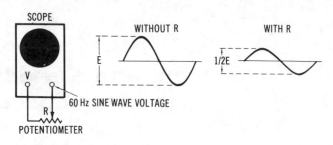

(A) Using the front-panel voltage source.

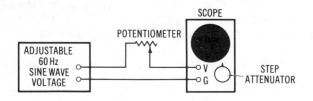

(B) Using an external voltage source.

Fig. 8-1. Checking the input resistance of an oscilloscope.

Next, to measure the vertical-input capacitance of the oscilloscope, a 100-kHz voltage is applied to the vertical-input terminal of the scope, as shown in Fig. 8-2. The amount of vertical deflection is first noted with no series capacitance in the circuit. Then, a trimmer capacitor is used to insert sufficient series capacitance so that the amount of vertical deflection is reduced to one-half. Then, the trimmer capacitor is disconnected and its capacitance value is measured with an accurate capacitance bridge. This value is equal to the vertical-input capacitance of the oscilloscope. *Particular attention should be given to the attenuator input resistance and capacitance on any step that does not have a correct attenuation factor with respect to the adjacent attenuator steps. Although it may be found that the adjustment of the compensating capacitor is incorrect on the out-of-limits step, it is also quite possible that capacitor leakage, an off-value resistor, or a switch defect will be found.*

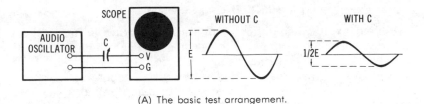

(A) The basic test arrangement.

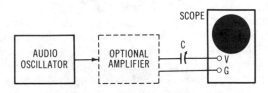

(B) An amplifier will step up the test voltage.

Fig. 8-2. Checking the input capacitance of an oscilloscope.

The *sensitivity* of an oscilloscope can be easily measured with the arrangement shown in Fig. 8-3. Sensitivity should be measured only after the condition of the vertical step attenuator has been verified. The sensitivity check can be made on any step of the vertical attenuator. With reference to Fig. 8-3, the output from the audio oscillator is adjusted for a vertical deflection of 1 inch (or 1 cm, or 1 division), depending upon the unit rating of the oscilloscope. Then, the vertical-input voltage is measured with a DVM or other accurate ac voltmeter. Suppose that this measurement is made on the × 100 step of the attenuator (with the fine vertical-gain control set to maximum). Then, the sensitivity of the oscilloscope will be 0.01 of the measured voltage. For example, if the DVM indicates 2 volts rms, the sensitivity of the oscilloscope is 20 mV (per inch, cm, or division, as the case may be). The *horizontal-amplifier sensitivity* can be checked in the same way.

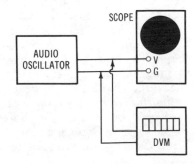

Fig. 8-3. Measurement of the vertical channel sensitivity of an oscilloscope.

FREQUENCY RESPONSE

A typical vertical-amplifier configuration is shown in Fig. 5-3. To check the frequency response of such a vertical amplifier, a test arrangement like the one shown in Fig. 8-4 is utilized. It is advantageous to use a calibrated lab-type signal generator. This type of generator has a good sine-wave waveform and provides a precise level of output voltage over its entire range. A vertical amplifier is usually rated for a frequency response to its − 3 dB point; this is 70.7% of its maximum voltage point. Many service-type scopes have a frequency response to 4 or 5 MHz; lab-type scopes have a frequency response to 15 MHz, or considerably more. Note that *horizontal-amplifier frequency response* can be checked in the same way; the only difference is that horizontal deflection is observed, instead of vertical deflection. Horizontal amplifiers typically have less sensitivity than vertical amplifiers, less frequency response, and a lower input impedance. For example, a typical horizontal amplifier is rated for a sensitivity of 0.025 V/cm, a frequency response to 500 kHz, and an input impedance of 100,000 ohms.

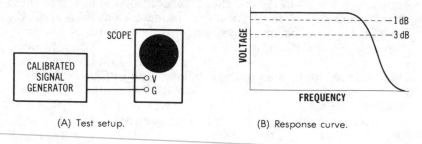

(A) Test setup.　　　　　　　　　(B) Response curve.

Fig. 8-4. Checking the frequency response of an oscilloscope.

SQUARE-WAVE RESPONSE

Oscilloscopes with triggered sweeps and calibrated time bases are generally rated for square-wave response in terms of rise time. For example, a typical service-type scope is rated for 50 nanoseconds (ns) rise time. With reference to Fig. 8-5, rise time is checked by applying the output from a square-wave generator to the vertical-input terminals of the scope. The leading edge of the displayed square wave is then expanded horizontally until the rise time can be properly measured. Note that *subnormal rise time* is usually associated with *subnormal frequency response* in the vertical amplifier. This condition is likely be *caused* by defective high-frequency compensating capacitors, such as capacitors C3 and C6 that are shown in Fig. 5-3. Subnormal rise time can also result from the replacement of transistors

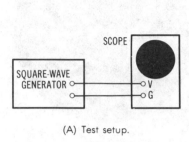

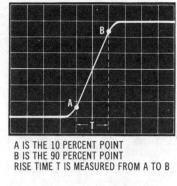

A IS THE 10 PERCENT POINT
B IS THE 90 PERCENT POINT
RISE TIME T IS MEASURED FROM A TO B

(A) Test setup.

(B) Response curve.

Fig. 8-5. Checking the vertical-amplifier square-wave response of an oscillator.

with types that have less high-frequency capability than the original transistors. As noted previously, tests of vertical-amplifier characteristics are made with the assumption that the vertical step attenuator is normal; otherwise, an attenuator malfunction would be charged to the vertical amplifier.

CHECK OF DEFLECTION LINEARITY

Although oscilloscopes are not always rated for deflection linearity, this is an important feature. For example, in hi-fi stereo troubleshooting, hi-fi units cannot be properly tested for distortion unless the oscilloscope has a high deflection linearity. A preliminary check of deflection linearity can be made as depicted in Fig. 8-6. The output from an audio generator is applied to both the vertical-input and the horizontal-input terminals of the scope. If the deflection linearity is good, a straight diagonal line is displayed on the scope screen. On the other hand, departures from linearity show up as a curvature in the display. When nonlinearity is found, it is next necessary to determine whether the trouble is in the vertical amplifier, in the horizontal amplifier, or both.

To determine which amplifier is operating nonlinearly, apply a signal from the audio generator to the vertical-input terminal, and reduce the horizontal deflection to zero in order to obtain a vertical-line display on the scope screen. Then, advance the step attenuator for full-screen vertical deflection—this might occur on the "5" step of the vertical attenuator, as exemplified in Fig. 8-7. Then, reduce the attenuator setting to "10." If the vertical amplifier is operating linearly, the trace height will fall to one-half its original

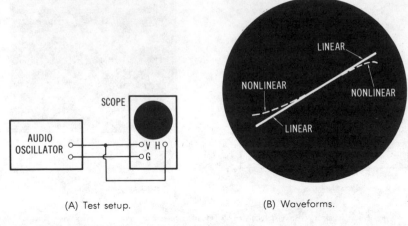

(A) Test setup.   (B) Waveforms.

Fig. 8-6. Checking the deflection linearity of an oscilloscope.

height. Reduce the attenuator setting to "20." If the vertical amplifier is operating linearly, the trace height will again be halved. If the vertical amplifier is found to be operating linearly, then the trouble will be found in the horizontal amplifier. Amplifier nonlinearity is most likely to be caused by marginal transistors; a collector-junction leakage is a common culprit. However, off-value bias voltages or a low collector-supply voltage can also be responsible.

CHECK OF TIME-BASE CALIBRATION

A check of time-base calibration can easily be made using a lab-type signal generator, as illustrated in Fig. 8-8. In this example, the time/division switch is set to 0.5 millisecond. In turn, a 1-kHz sine wave

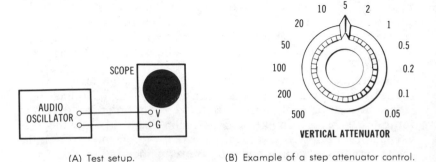

(A) Test setup.   (B) Example of a step attenuator control.

Fig. 8-7. Checking an oscilloscope for vertical-deflection linearity.

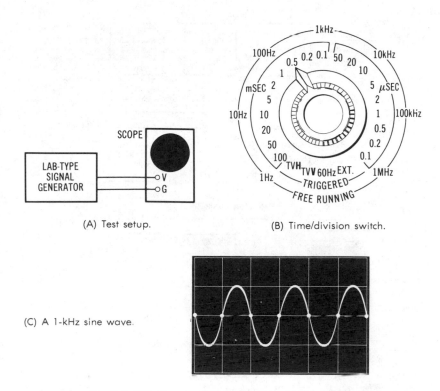

(A) Test setup.

(B) Time/division switch.

(C) A 1-kHz sine wave.

Fig. 8-0. Checking an oscilloscope's time-base calibration.

is normally displayed as illustrated in the diagram of Fig. 8-8C. (However, if the time/division switch were set to 0.5 µs, a 1-MHz sine wave would be displayed.) A schematic for a triggered time-base configuration is shown in Fig. 3-16. The sweep speed is determined by the values of the capacitors shown in the Q322 circuitry; note that the base of transistor Q322 is returned to potentiometer R330, the sweep-calibration maintenance control. In case the time-base calibration is incorrect, a simple readjustment of the maintenance control may be all that is required. On the other hand, if a time-constant capacitor is leaky or otherwise defective, it will be impossible to bring the time base into full calibration until the defective capacitor is replaced. Note that marginal transistors can also cause time-base calibration errors.

Oscilloscope manufacturers often provide troubleshooting charts for their instruments. For example, a block diagram of an oscilloscope is shown in Fig. 8-9. It is used as a troubleshooting reference. The oscilloscope is divided into nine sections. The transistors associated with each section are indicated. Operating controls are shown with

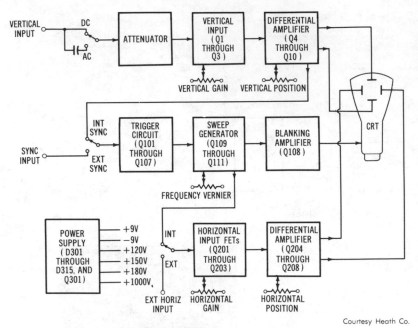

Fig. 8-9. Block diagram of an oscilloscope.

their related sections. Power-supply voltages are also indicated. *The first rule in an oscilloscope troubleshooting procedure is to measure the power-supply voltages. The second rule in oscilloscope troubleshooting is that a good scope must be used to signal trace the circuits in a defective scope.* Some trouble symptoms and possible areas of malfunction for the scope diagrammed in Fig. 8-9 are listed in Table 8-1. When a trouble symptom is tracked down to a particular stage, the dc voltages should be measured at the transistor terminals in order to pinpoint the defective device or component. Sometimes resistance measurements must also be made. Note that a high-power/low-power ohmmeter is most useful in such applications, because a low-power ohmmeter can be applied as if the transistor junctions were open circuits. (This rule does not apply to defective transistors, however.) To repeat an important word of caution, *an oscilloscope operates at high voltage.* The crt heater and cathode circuits are often 1000 volts, or more, above ground. Also, dangerously high voltages can appear at unexpected points in defective circuitry.

As a practical note, servicing data may not always be available for a particular oscilloscope. In such a case, it is often possible to find a similar oscilloscope in a nearby service shop. Comparison tests can then be made between the defective scope and the normal scope. This is sometimes an advisable procedure, even when servicing data are

Table 8-1. Troubleshooting Chart

Difficulty	Possible Area of Trouble
Neither pilot lamp nor crt filaments light.	1. Fuse blown. 2. On-off switch. 3. No ac power from outlet.
Pilot lamp lights, crt filament does not light.	1. Power transformer. 2. Crt.
No spot or trace on crt.	1. Positioning or intensity controls improperly adjusted. 2. High-voltage power supply. 3. Crt.
Dot cannot be centered vertically.	1. Vertical position control and associated circuit.
Dot cannot be centered horizontally.	1. Horizontal position control and associated circuit.
No vertical deflection.	1. Vertical amplifier.
No horizontal deflection.	1. Horizontal amplifier.
Poor focus. Astigmatism.	1. Crt. 2. Focus control. 3. Astigmatism control. 4. Resistors R412, R413, R414, and R303. 5. Incorrect crt anode voltages.
Trace acts erratic when the window is touched.	1. Clean the window with detergent to eliminate static charge.
Cannot synchronize input signal with sweep generator frequency.	1. Sync switch in the EXT position. 2. Control R103 misadjusted.
No retrace blanking or poor retrace blanking.	1. Transistor Q108. 2. Diode AZ101.
Pilot lamp changes intensity from bright to dim.	1. This is normal operation.

Courtesy, Heath Co.

available for the defective scope. In other words, servicing data are seldom completely comprehensive, whereas, a comparison scope provides ready data concerning the operating characteristics, voltage values, resistance values, current values, and electrical changes in response to control variations. In any case, the troubleshooter's prime assets are a good knowledge of electrical circuit action, the ability to reason logically, and experience in the use of electrical measuring instruments.

A generalized oscilloscope troubleshooting chart is shown in Table 8-2. These possible remedies apply to almost any oscilloscope. The chart should be used in conjunction with a circuit diagram for the particular oscilloscope, so that the trouble area can be "tracked down" on the scope chassis. It is good practice to start by looking for obvious

Table 8-2. Generalized Troubleshooting Chart

Trouble	Possible Remedy
Pilot light will not operate.	Check ac power cable, fuse, pilot light, or power transformer.
No spot on cathode-ray tube.	Check horizontal or vertical positioning control, high-voltage, vertical, and horizontal amplifier collector load resistor.
No control of focus or intensity.	Check cathode-ray tube socket, high-voltage bleeder open, focus control, and intensity control.
No control of vertical positioning.	Check vertical amplifier, vertical positioning control, low-voltage power supply, and vertical amplifier transistors.
No control of horizontal positioning.	Check horizontal amplifier, horizontal positioning control, low-voltage power supply, and horizontal amplifier transistors.
No vertical deflection.	Check vertical amplifier transistors, low-voltage power supply, input leads, and vertical attenuator switch.
Vertical deflection but poor frequency response.	Check input- and output-stage frequency compensators.
No horizontal sweep on some settings.	Check coarse-frequency range switch, and range capacitors.
No horizontal sweep on any setting.	Check sweep-circuit oscillator transistor.
Horizontal deflection but poor frequency response.	Check horizontal amplifier tubes, attenuator switch, low-voltage power supply, input leads, and output frequency compensators.
No sync.	Check sync selector switch, sync injector tube, and locking control.
No blanking.	Check coupling capacitor.

defects such as components or devices that have broken loose from one end, scorched resistors or other evidence of overheating, leaking electrolytic capacitors, abnormal power-transformer hum (evidence of excessive current drain), sharp odors, arcs, or smoke. Sometimes switch contacts are corroded or loose. If a chassis is dusty or grimy and has collected lint, it should be cleaned before troubleshooting is started. In some cases, the trouble will be found before the cleaning job is completed. After the malfunction has been corrected, it should not be assumed that the instrument is now working correctly on all functions—each specification should be verified, as explained in the first part of the chapter.

As noted in Table 8-1, astigmatism (Fig. 8-10) results from incorrect anode voltages. Some oscilloscopes provide an astigmatism control, as exemplified in Fig. 8-11. Note that the astigmatism control and the focus control adjust the potential on two separate anodes. Both adjustments are comparatively critical. If an oscilloscope does not provide an astigmatism control, the crt potentials should be accurately measured in case of astigmatism. The measured values should agree, within reasonable tolerance, with the values specified in the oscilloscope manual. Note that unbalanced outputs from either the vertical amplifier or from the horizontal amplifier can also cause

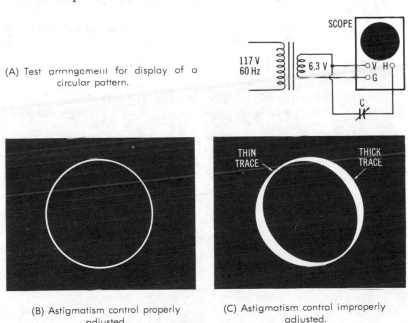

(A) Test arrangement for display of a circular pattern.

(B) Astigmatism control properly adjusted.

(C) Astigmatism control improperly adjusted.

Fig. 8-10. Focus and astigmatism.

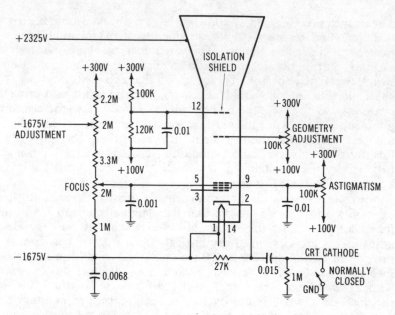

Fig. 8-11. Example of crt circuitry that contains an astigmatism control.

astigmatism. Therefore, if the crt potentials check out correctly, the output amplifiers may be at fault. The geometry control in Fig. 8-11 is found on comparatively sophisticated scopes. It is adjusted to obtain optimum deflection linearity at the extreme ends of the trace. Note also that a high-voltage adjustment control is provided in this example; it is adjusted to obtain a crt cathode potential of -1675 volts. Keep in mind that all of the crt potentials interact to some extent.

CHAPTER 9

Frequency and Phase Measurements

BASIC METHODS OF FREQUENCY AND PHASE MEASUREMENT

Frequency measurement with a triggered-sweep oscilloscope that has a calibrated time base is depicted in Fig. 9-1. If the sweep-speed control is set to 100 μs per division, and a complete cycle occupies four divisions, the period of the waveform is 400 μs and its frequency is 2500 Hz. However, if a scope with a free-running sweep is used, a method of frequency measurement that uses the setup depicted in Fig. 9-2 may be utilized. This is called Z-axis or intensity modulation of the displayed waveform. The generator is carefully adjusted to make the blanked intervals "stand still" on the screen pattern. Each interval represents a peak in the modulating waveform. Thus, if the generator frequency is 6 kHz in the example shown, the frequency of the displayed waveform is

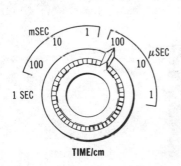

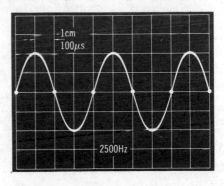

Fig. 9-1. Frequency measurement can be made with a triggered-sweep scope that has a calibrated time base.

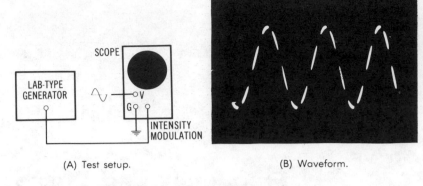

<table>
<tr><td>(A) Test setup.</td><td>(B) Waveform.</td></tr>
</table>

Fig. 9-2. Frequency measurement by Z-axis modulation.

1 kHz. (There are six blanked intervals in one cycle of the screen pattern.) Note that if a dual-trace scope is employed, a known frequency can be compared with an unknown frequency, as shown in Fig. 9-3. Note also that this method indicates the phase difference between two waveforms that have the same frequency.

Distinction should be made between measurement applications and indicator applications. For example, the procedure shown in Fig. 9-1 is a *measurement technique*, inasmuch as the oscilloscope serves as a measuring instrument. On the other hand, the procedure depicted in Fig. 9-2 employs the scope only as an *indicator*—it is the lab-type generator which serves as the measuring instrument. In the example of Fig. 9-3, the dual-trace scope is being used as an *indicator*—the known frequency provides the basis for measurement. Note, however, that if the scope has a calibrated time base, it can be operated as a measuring instrument in this application. The phase difference between two waveforms that have the same frequency can be measured with a single-trace scope, using its external-sync function, as depicted in Fig. 9-4. In this application, the scope operates as an indicator—the

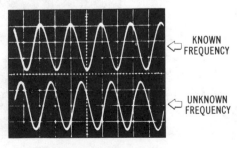

KNOWN FREQUENCY

UNKNOWN FREQUENCY

Fig. 9-3. A known frequency can be compared to an unknown frequency when using a dual-trace scope.

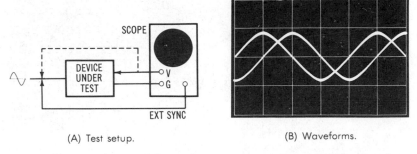

(A) Test setup.   (B) Waveforms.

Fig. 9-4. Checking the phase difference between two waveforms using the external-sync function of a scope.

reference waveform itself serves as the unit of phase measurement. Note in passing that two waveforms that have different frequencies can have a phase difference with *respect to a specific time,* as exemplified in Fig. 9-5. (The phase relation at t_0 in Fig. 9-5B corresponds to the peak of the modulation signal envelope in Fig. 9-5A.)

Another basic type of phase check is illustrated in Fig. 9-6. This is an example of complex waveforms wherein each waveform consists of "sine waves" and blanking pulses. The "sine-wave" components have the same frequency, although their relative phases are different. Phase is defined in terms of nulls—in other words, points along the horizontal axis at which the "sine pulses" fall to zero (or practically zero) amplitude. Ten intervals with 30° separation are observed between successive blanking pulses. In turn, relative phases for normal operation are defined in terms of nulls at particular intervals—the R-Y waveform nulls on the sixth interval; the B-Y waveform nulls on the third and ninth intervals; the G-Y waveform nulls on the first and seventh intervals.

LISSAJOUS FIGURES

There are many types of Lissajous figures that can be displayed with an oscilloscope. Vectorgrams, as exemplified in Fig. 9-7, are familiar to color-tv technicians. Fig. 9-8 shows how basic Lissajous figures are developed. The internal sweep system of the oscilloscope is turned off, and two different sine-wave signals are fed to the inputs of the scope, one to the vertical-input terminal and the other to the horizontal-input terminal of the scope. The resulting display is a closed-loop waveform named after the French scientist Lissajous, who first showed how these closed-loop waveforms could be developed.

If the frequency of one of the generators is known, the other, or unknown, frequency can be determined by proper interpretation of

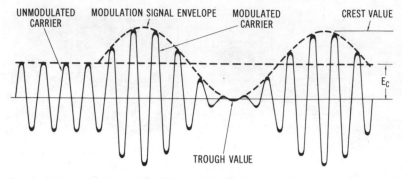

(A) Signal waveform.

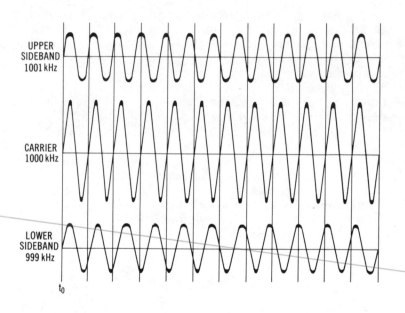

(B) Separate waveforms.

Fig. 9-5. Two waveforms that have a phase difference that is defined with respect to time.

the Lissajous figure generated. The known frequency signal is usually connected to the horizontal input of the scope, and the unknown signal is connected to the vertical input. Simple frequency ratios result in waveforms that are easy to interpret. Therefore, the known frequency generator should be adjusted to obtain such a waveform, if possible. When the two signals have a ratio that can be expressed by whole

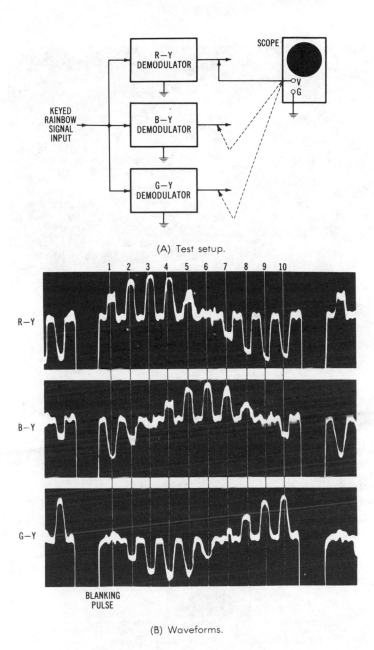

(A) Test setup.

(B) Waveforms.

Fig. 9-6. Phase checks of R-Y, B-Y, and G-Y waveforms.

149

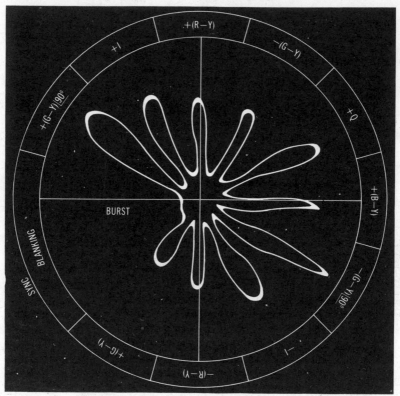

Fig. 9-7. Display of a typical vectorgram.

Courtesy Sencore, Inc.

numbers, such as 2:3, 3:4, 4:1, etc., the waveform will appear to be stationary on the scope screen. If one generator is adjusted to give a signal ratio not quite expressible by two whole numbers, the pattern will seem to rotate slowly in one direction or the other. This rotation appears to take place in three dimensions, a fact that can sometimes be used to advantage in counting the loops of the pattern.

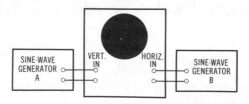

Fig. 9-8. A setup for generating Lissajous figures.

150

Fig. 9-9 shows the pattern resulting from a 2 to 1 frequency ratio and illustrates the method used for calculating the frequency ratios of other patterns. Note that horizontal line AB is drawn touching the pattern at two points, and that vertical line AC is drawn touching the pattern at one point. The points of contact on line AB are caused by vertical excursions of the oscilloscope beam and are therefore related to the frequency of the signal applied to the vertical input of the scope. The points of contact on line AC (only one in this instance) are caused by horizontal excursions of the beam and are therefore related to the frequency of the signal applied to the horizontal input. From the pattern shown in Fig. 9-9, the oscilloscope beam makes two vertical excursions while making one horizontal excursion; therefore, the vertical input signal has twice the frequency of the horizontal input signal. Three points of contact to line AB and one to line AC would indicate that the vertical input-signal frequency was three times that of the horizontal input signal. Three points of contact on line AB and two on line AC would indicate that the vertical input-signal frequency was 3:2 that of the horizontal input signal, and so on.

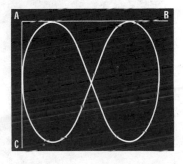

Fig. 9-9. A method for calculating the frequency ratio indicated by a Lissajous figure.

Once the frequency ratio indicated by the Lissajous figure has been determined, the unknown frequency can be found by multiplying the known frequency by this ratio. A few of the simpler ratios are shown in Fig. 9-10. At "A," a 3 to 2 ratio is shown. If this pattern is permitted to rotate, it will appear as shown at B or C. The pattern at C could lead to some error in interpretation. The point to remember is that the pattern should consist of closed loops rather than abruptly terminated single lines as it seems to do in Fig. 9-10C. The pattern of Fig. 9-10D indicates a 5 to 3 frequency ratio, while that of Fig. 9-10E indicates a 4 to 5 ratio.

A continually shifting Lissajous pattern results when the phase relationship between the two input signals is constantly changing. The more complex the pattern (resulting from a frequency ratio having large numbers, for example, 17:13), the harder it is to interpret. The task is made even more difficult by a shifting pattern. It is better, then,

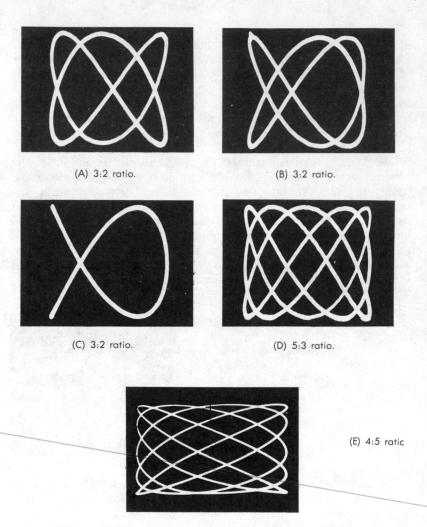

(A) 3:2 ratio. (B) 3:2 ratio.

(C) 3:2 ratio. (D) 5:3 ratio.

(E) 4:5 ratic

Fig. 9-10. Patterns used to determine frequency ratios between pairs of input signals.

to simplify the ratio, if possible, by changing the known frequency. If this is not practical, other methods of frequency determination may work better.

PHASE COMPARISON AND MEASUREMENT

The oscilloscope can be used to make phase comparisons between two signals of the same frequency. The two signals are fed to the vertical and horizontal inputs in the customary manner to develop a

Lissajous pattern. The phase relationship between the two signals can be determined by proper interpretation of this pattern.

The following discussion and waveforms are based on these arbitrarily chosen phase relationships.

1. The horizontal signal is considered to be leading the vertical signal by the specified amounts.
2. The same number of phase reversals take place in the vertical as in the horizontal amplifiers of the oscilloscope, if amplifiers are used.
3. A positive-going signal applied to the vertical input causes an upward deflection, and a positive-going signal applied to the horizontal input causes deflection to the right.

Fig. 9-11 illustrates how two sine-wave signals applied to the vertical and horizontal deflection plates develop a pattern indicative of the

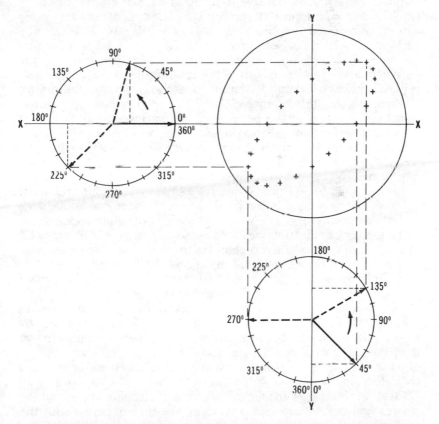

Fig. 9-11. Graphic illustration of the manner in which two sine-wave signals of identical frequency, but different phase, develop a phase indication on an oscilloscope.

153

phase relationship between the two signals. The large circle represents the face of the scope with X and Y axes drawn through its center. Voltage applied to the horizontal-deflection plates moves the beam along the X-axis, right or left of center, depending on the polarity of the voltage. Voltages applied to the vertical-deflection plates produce beam movement along the Y-axis.

The two smaller circles represent sine-wave generators for the vertical and horizontal signals and are divided from zero to 360° in steps of 22.5°. The radial arrows are vector representations of the maximum signal voltage of each generator and are made equal to each other for simpler illustration. The deflection factors of the vertical and horizontal plates of the scope are assumed to be equal for the same reason.

In the left-hand circle, a perpendicular to the X-axis from the point of the radial arrow will represent the instantaneous magnitude of the sine-wave voltage applied to the vertical-deflection plates. This value can be transferred graphically to the scope diagram to locate the vertical position of the beam trace at that instant. Other times during the cycle are indicated by dotted arrows. In like manner, the perpendicular to the Y-axis from the arrow point in the lower circle represents the horizontal-deflection voltage at any particular instant. This value is transferred graphically to the scope diagram to locate the horizontal position of the beam trace at that instant.

For this illustration, the two sine-wave vectors have been chosen so that the horizontal vector is 45° ahead of the vertical vector. Since the frequencies of both signals are the same, this 45° difference is maintained throughout the complete cycle. The beam position on the scope is plotted for every 22.5° interval of the cycle, and the resultant graph gives a very good indication of the scope waveform obtained for a phase difference of 45° between signals. The waveform is an ellipse and is the same as that obtained for a phase difference of 360° minus 45° (315°), as will be seen if we consider the horizontal vector as zero reference, with the vertical vector lagging the horizontal.

Any phase difference can be plotted in this manner, and a complete series would show that the pattern is either a straight line, a circle, or an ellipse. Straight lines occur at 0° and 180° phase difference; circles occur at 90° and 270°. All other values of phase difference are shown by ellipses. These ellipses become narrow and approach a straight line at 0° or 180° or become broader and approach a circle at 90° and 270°. A number of these patterns are shown in Fig. 9-12. Phase differences are indicated for intervals of 45°, from 0° to 360°.

Fig. 9-11 shows how the technician can plot any phase difference he desires and get an accurate waveform like that obtained with the oscilloscope. In practical cases, he may be interested in somewhat the reverse effect. That is, he may have a Lissajous pattern and desire to

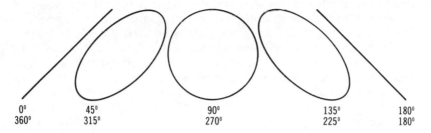

| 0° | 45° | 90° | 135° | 180° |
| 360° | 315° | 270° | 225° | 180° |

Fig. 9-12. Phase indications from 0° to 360° at intervals of 45°.

know the phase difference it represents. Fig. 9-13 shows how an unknown phase angle can be calculated from such a pattern.

It is convenient to consider all the phase angles from 0° to 90° as basic and to calculate all the others through them. Thus, referring to Fig. 9-12 again, at 0° we start with a straight line (a closed loop seen on edge), and go through an infinite series of ellipses, finally arriving at a circle for 90°. The same series of ellipses, but slanting in the opposite direction, covers the range from 90° to 180°. The range from 180° to 360° duplicates both series, but in reverse order. Before making the measurements of Fig. 9-13, the vertical and horizontal gains should be adjusted to be as nearly equal as possible; otherwise, a perfect circle will not be obtained at 90°. The accuracy of this method will not be impaired, however, if the two gains are not exactly equal.

The ellipse should be positioned so that its center coincides with the intersection of the graph lines of the scope calibration screen as shown in Fig. 9-13. If the distances C and D are measured and substituted in the formula given, we obtain the sine of the phase angle. The phase angle can then be found by locating this value in a table of sines.

Phase angles between 90° and 180° (their ellipses slant from lower

Fig. 9-13. A method of calculating the phase angle represented by a 1-to-1 Lissajous figure.

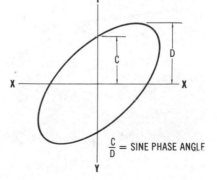

$$\frac{C}{D} = \text{SINE PHASE ANGLE}$$

right to upper left) are found in the following manner. Find the value of the sine by the preceding method. Locate this value and corresponding angle in the sine table. An angle between 0° and 90° will be given, which should then be subtracted from 180° for the final correct value.

Notice in Fig. 9-12 that each waveform is labeled with two values of phase angle. When the technician views the waveform, there is no indication to the eye which phase angle is the correct one. However, there is a difference in how the waveform is developed. In one instance, the beam is traveling clockwise around the waveform and, in the other, the beam is traveling counterclockwise. Under the conditions stated in the second paragraph of this discussion of phase comparisons, the beam travels counterclockwise for the 45°, 90°, and 135° waveforms of Fig. 9-12, and clockwise for the 315°, 270°, and 225° waveforms.

The direction of beam travel can be determined by at least two methods.

1. By changing the phase of one signal in a known direction.

2. By intensity modulation of the beam with a suitable marker.

As an example of the first method, suppose we desire to know whether an ellipse like the first, shown in Fig. 9-12, represents a 45° or 315° phase difference. We know the horizontal signal is leading the vertical signal by one or the other of these amounts. Suppose we increase the lead of the horizontal signal. If the waveform changes in the direction of the circle, it originally was a 45° waveform, but if the waveform changes toward a straight line, it was the 315° waveform.

Fig. 9-14 is one example of the type of signal that can be used to mark a pattern so that the direction of beam travel can be determined. This illustration shows just the positive peaks of the signal since they are responsible for the visible indication on the waveform. This type of signal is effective for this purpose because, as it is made more and more

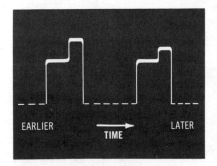
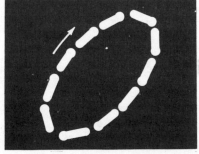

Fig. 9-14. Intensity marker signal at A produces markers as at B. Beam rotation is as indicated by arrow.

positive, the beam trace gets brighter and has a tendency to enlarge. The signal at A in Fig. 9-14 will then result in a marked trace like the one at B. Thus, a trace resembling a series of arrows pointing in the direction of beam travel is obtained. The frequency of the marking signal is unimportant except it should be greater than the vertical input signal and a whole number multiple of this frequency. The intensity adjustment of the oscilloscope should be reduced to nearly minimum intensity for a more effective marker. If a marking signal of the type shown at A in Fig. 9-14 is not available, others may be used. A sawtooth signal like the one developed by the scope generator will produce a narrow wedge-shaped marker. The peak of the sawtooth will correspond to the wide end of the wedge. Which portion of the sawtooth signal occurs later in time must be known to determine the direction of beam travel.

CHAPTER 10

Amplifier Testing

OSCILLOSCOPE APPLICATIONS IN AMPLIFIER TESTING

Various amplifier characteristics are of concern to the technician, such as frequency response, distortion, noise output, phase-shift, tone-control action, stability, equalization pattern, power output, and so on. Noise is a common problem in hi-fi stereo system operation, and it can have various causes. For example, worn discs and deteriorated tapes are a source of noise apart from the amplifier system. Tuners may have a poor signal-to-noise ratio, apart from the amplifier system. When noise originates in a preamplifier or in a power amplifier, it is often the result of collector-junction leakage in a transistor (Fig. 10-1). However, noise can also originate in resistors. The resistors in a preamplifier section are ready suspects because any noise that they generate is then amplified by the following stages. Note that metal-film resistors have an inherently lower noise level than composition resistors. Noise can be signal-traced by an oscilloscope to its point of origin.

A defective negative-feedback circuit can result in a higher noise output level, because negative feedback reduces noise that originates in the amplifier, as depicted in Fig. 10-2. Observe that negative feedback operates on the principle of predistortion. In other words,

Fig. 10-1. A leaky collector junction often generates noise.

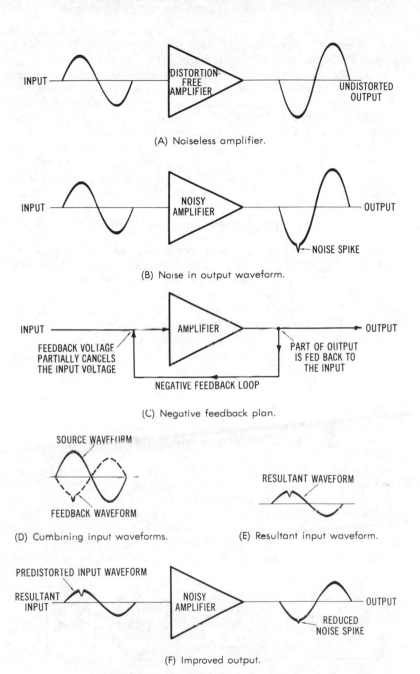

(A) Noiseless amplifier.

(B) Noise in output waveform.

(C) Negative feedback plan.

(D) Combining input waveforms.

(E) Resultant input waveform.

(F) Improved output.

Fig. 10-2. Negative feedback reduces noise that originates in the amplifier.

when a noise spike appears in the output of the amplifier, this spike is fed back in opposing phase to the input. The combining of the input waveform and the feedback waveform produces a resultant predistorted waveform in which the noise spike is inverted. Therefore, the noise spike that is generated within the amplifier becomes partially cancelled by the inverted noise spike from the feedback loop. In turn, the output waveform from the amplifier displays a greatly reduced noise spike. The two principal types of negative feedback are voltage feedback, as exemplified in Fig. 10-2, and current feedback as produced by an unbypassed emitter resistor. Either type of negative feedback helps to reduce noise that originates within the amplifier.

EVALUATING DISTORTION PRODUCTS

With reference to Fig. 10-3, a scope is very helpful in evaluating distortion products. A harmonic distortion meter will measure the percentage of total harmonic distortion, but it cannot show whether the distortion is being caused by a second harmonic, by a third harmonic, by a fourth harmonic, by hum voltage, or by noise voltages. However, if a scope is connected at the output of the harmonic distortion meter, the nature of the distortion products can be determined. For example, the scope might show that the distortion products consist principally of noise voltages, or that low-order or high-order harmonics are present (harmonic frequencies can be measured with a calibrated time base). Or, the scope might show that the distortion consists chiefly of 60-Hz or 120-Hz hum voltage. After the technician identifies the nature of the distortion products, he can zero in on the defective area rapidly and with disregard of "impossibles."

Note that the oscilloscope in Fig. 10-3 shows the harmonics that may be present in the output waveform because the test frequency applied by the audio generator is filterd out by means of a notch filter in the harmonic distortion meter. Fig. 10-4 shows how a distorted resultant output waveform is produced by a second harmonic or by a third harmonic. *Even harmonics* tend to be produced by compression or clipping of a single peak in the output waveform, whereas *odd*

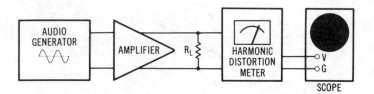

Fig. 10-3. Oscilloscope displays amplifier distortion products.

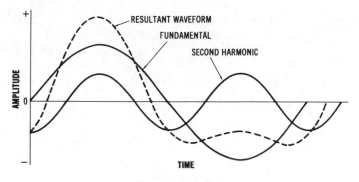

(A) Scope displays second harmonic.

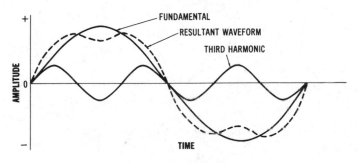

(B) Scope displays third harmonic

Fig. 10-4. Basic second- and third-harmonic distortions.

harmonics tend to be produced by symmetrical peak compression or clipping. However, it is quite common for an amplifier to develop both even and odd distortion products. For example, Fig. 10-5 shows how a distorted resultant output is produced by a combination of second-harmonic and third-harmonic products. Observe that second-harmonic distortion results in an unsymmetrical output waveform, whereas third-harmonic distortion results in a symmetrical output waveform. When a second harmonic beats with a third harmonic, the resultant waveform is unsymmetrical due to the effect of the even harmonic.

SQUARE-WAVE FREQUENCY-RESPONSE TEST

The merit of the square-wave test to indicate an amplifier frequency response is based on the fact that a square wave represents a great many frequencies other than its own fundamental frequency. Fourier analysis has shown that a square wave can be built up from many sine

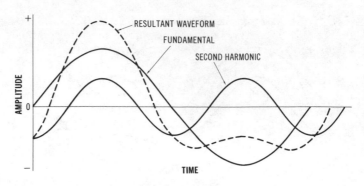

(A) Result of second-harmonic distortion.

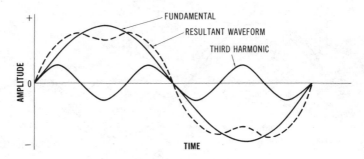

(B) Result of third-harmonic distortion.

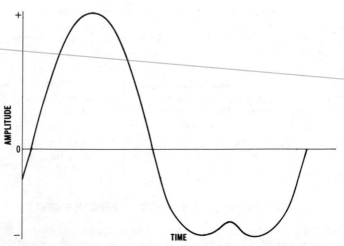

(C) Result of combined second- and third-harmonic distortion.

Fig. 10-5. Second-harmonic and third-harmonic distortion.

waves of different amplitudes and frequencies. The fundamental frequency will have the greatest amplitude, and it is combined with odd-numbered harmonics that decrease in amplitude as the order of the harmonics increases. Thus, a square-wave signal of a certain fundamental frequency applied to an amplifier is a more extensive test of the amplifier response than is a sine-wave signal of the same frequency. The square-wave test for frequency response consists of applying the square-wave signals of various fundamental frequencies to the amplifier and observing the output on an oscilloscope. Any change in the shape of the square wave can then be interpreted in terms of frequency response of the amplifier.

This test is not the type that will furnish information for plotting a response curve. That is, the exact ratio between the amplification factors at various frequencies will not be determined. Rather, a quick overall picture of the frequency response will be gained. Before a test, the quality of the square-wave signal put out by the square-wave generator should be checked. The signal should be fed directly to the oscilloscope input and checked for flatness and sharp corners. A slight imperfection can be tolerated and allowed for when the amplifier response is interpreted. The oscilloscope amplifiers should have response characteristics as good as, or better than, the amplifier being tested; otherwise, they become a limiting factor in the test. If the output waveform from the square-wave generator appears good for some settings of the scope attenuator switch and peaked or rounded at another setting, the attenuator adjustment for that switch position should be checked and readjusted if necessary.

It is customary to assume that a well-reproduced square wave of any particular frequency indicates good response for all frequencies from one-tenth to ten times that frequency. Since this frequency range of 100 to 1 is far short of the range of most present-day amplifiers, the square-wave generator must be reset several times to make a complete test. Usually the extreme low and high ends of the amplifier response will interest the technician more than the midrange. They determine the width of the response range and present more design difficulties to the manufacturer.

An acceptable test is to run the square-wave frequency low enough and high enough until the response falls off at these extremes and then tune the generator through the entire range between, meanwhile watching for any peculiarities.

The technician can acquaint himself with some of the square-wave response curves obtainable if he checks the action of the bass and treble controls during a test. Some of the following waveforms were obtained in this manner. Fig. 10-6 resulted from applying a 1-kHz signal directly to the vertical input of the oscilloscope. The tops and bottoms of the square wave are straight and level, with no evidence of

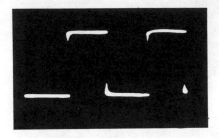

Fig. 10-6. A 1-kHz square wave
applied directly to the
oscilloscope input.

overshoot or ringing. Fig. 10-7 shows the same signal applied to the tuner input of a preamplifier of the type that is used between a tuner or phonograph pickup and an amplifier. The bass and treble controls were adjusted for a response that approaches that in Fig. 10-6 as nearly as possible. A slight peak remains at the leading edge of the square wave, and this tendency can be noticed in some of the illustrations which follow. A sine-wave frequency check shows the amplifier to be flat over the audio range. The most likely explanation of the peak is that it is due to some overshoot in the preamplifier circuits.

When the treble control is advanced, the leading edge of each cycle becomes even more peaked, as in Fig. 10-8. The peak appears for both positive and negative halves of the cycle. A logical conclusion from Fig. 10-8 is that excessive highs in an amplifier are indicated by peaked leading edges of the square-wave response, sloping back to the trailing edge. Conversely, the leading edges should be depressed below normal if the highs are attenuated, and this is exactly what happens.

When the treble control is kept normal and the bass control advanced, the waveform of Fig. 10-9 results. Trailing edges of each half cycle of the square wave are peaked, sloping gradually from the leading edges. When the bass control is set for attenuation, the waveform slopes in the opposite direction. It then resembles the waveform for treble emphasis. This is about what you might expect.

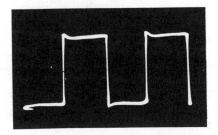

Fig. 10-7. Preamplifier response to
the 1-kHz square wave. Tone controls were adjusted for
flat response.

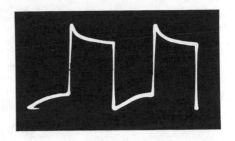

Fig. 10-8. Treble emphasis applied to a square-wave signal.

After all, it is a relative matter—reduce the bass or increase the treble—the results are similar.

Fig. 10-10 is the result of applying a 50-Hz square wave to the preamplifier with the tone controls set for flat response at 1 kHz. Leading edges are elevated and trailing edges are depressed—the marks of low-frequency attenuation and phase shift. The severity of a square-wave test over a sine-wave test is borne out by the fact that the preamplifier showed practically no attenuation to a 50-Hz sine-wave signal when compared to its 1000-Hz response, yet the 1000-Hz square wave of Fig. 10-6 was changed to that of Fig. 10-10 at 50 Hz.

When the square-wave generator is adjusted toward the high-frequency extreme, a point is reached where the preamplifier response begins to fall. This is evidenced by a gradual rounding of all corners of the square wave, as in Fig. 10-11, which shows the response to a 10 kHz signal. As the square-wave frequency is adjusted higher and higher, the corners are rounded more and more, and the square wave takes on the appearance of a sine wave.

THE SQUARE WAVE AS AN INSTABILITY CHECK

We have seen how the square-wave signal provides a quick check for the frequency-response range of an amplifier; it can also discover any

Fig. 10-9. Effect of bass boost on a square-wave signal.

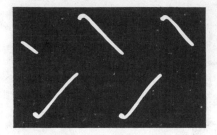

Fig. 10-10. A 50-Hz square wave shows low-frequency attenuation and phase shift.

tendency toward instability. The steep wavefront of the square wave can shock borderline cases into ringing or oscillation, as shown by Figs. 10-12 and 10-13. Ringing is just a form of oscillation that dies away quickly. In Fig. 10-12, about 3 cycles of oscillation are visible in each half cycle of square wave. Fig. 10-13 indicates a more unstable condition; the oscillations persist throughout the entire cycle, although they may not show in all parts of the reproduction. When this type of oscillation occurs in audio amplifiers, it is usually above the audio-frequency range and, therefore, will not be heard in itself, but it may react with the audible signal to cause distortion. It is almost certain to cause a lowering of the maximum power output.

Although an amplifier may be shock excited to the point of oscillation with a square-wave signal, it might not do so when a sine-wave signal is applied. The results of the square-wave test indicate, however, that some trouble might be expected on large audio signals of complex waveforms.

Video amplifiers in tv receivers can be tested with a square-wave signal in much the same manner as audio amplifiers. Interpretation of the resultant waveforms is similar to the examples just given. The range of frequencies is a little different with the tv receiver; it is shifted toward the higher frequencies. That is, it is usually not expected to go as low, but it does extend up to several megahertz. Here again, the

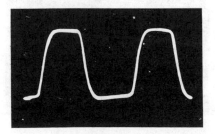

Fig. 10-11. High-frequency attenuation is shown by rounded corners of the square wave.

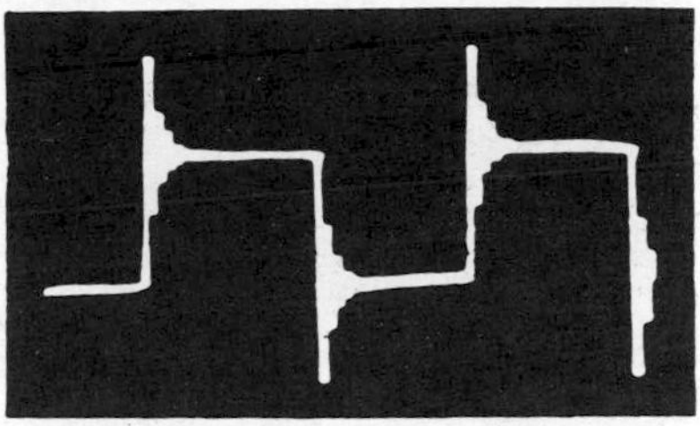

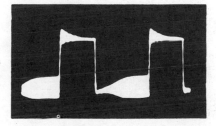

Fig. 10-12. Moderate ringing induced by application of square waves.

operator must be certain that the oscilloscope amplifiers do not become the limiting factor in response at these higher frequencies, or a misleading waveform will be obtained.

Phase shift and ringing are more evident in the tv picture than in an audio signal and will cause smearing and repeated outlines during reception of a regular broadcast signal. Incidentally, the picture tube can serve as a fair substitute for the scope in the square-wave test since the signal applied to the video amplifiers is fed directly to it. By turning the brightness up and down, the operator can get a good idea of the condition of the square-wave signal at the picture tube. For example, sharp corners on the waveform will result in sharp divisional edges between the light and dark bars on the screen; rounded corners will result in blended edges to the bars.

HUM PATTERNS

When hum voltage appears in a square-wave display, the hum voltage may or may not be synchronized with the square-wave frequency. Thus, the technician may observe either of the basic displays exemplified in Fig. 10-14. If the square-wave repetition rate is integrally related to the hum frequency, the result is to introduce a sinusoidal curvature into the waveform envelope. On the other hand, if the square-wave repetition rate is not integrally related to the hum frequency, the result is to produce a "thickening" along the top and bottom excursions of the square wave. In either case, the peak-to-peak hum voltage is easily measurable on the scope screen, as shown in Fig.

Fig. 10-13. Instability shown by continuous oscillation when square wave is applied.

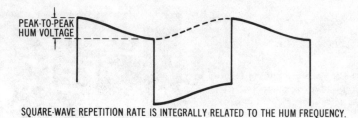

PEAK-TO-PEAK
HUM VOLTAGE

SQUARE-WAVE REPETITION RATE IS INTEGRALLY RELATED TO THE HUM FREQUENCY.
ELECTRON BEAM "FOLLOWS" THE HUM VOLTAGE.

(A) "Synchronous" hum voltage results in a sinusoidal envelope.

PEAK-TO-PEAK
HUM VOLTAGE

SQUARE-WAVE REPETITION RATE IS NOT INTEGRALLY RELATED TO THE HUM VOLTAGE.
ELECTION BEAM DOES NOT "FOLLOW" THE HUM VOLTAGE.

(B) "Nonsynchronous" hum voltage results in a thickening of top and bottom excursions.

Fig. 10-14. Basic displays of a square wave with hum voltage.

10-14. Note that the scope can be used to trace hum interference to its point of origin. Of course, if the hum voltage is entering via the power supply, the hum interference will be apparent at all stages in the amplifier system.

CHAPTER **11**

Servicing Digital Equipment

TYPICAL DIGITAL EQUIPMENT

A familiar example of digital equipment is a scanner-monitor radio receiver. This type of receiver continuously samples a group of channels for activity and when a signal output is detected from any channel, the scanning action stops, and the receiver locks on the active channel. At the end of station transmission, the scanning section resumes the sampling of the channels. A simplified block diagram for a 4-channel digital-logic section of a scanner-monitor receiver is shown in Fig. 11-1. Note that the *clock* generates a train of square waves or pulses. This pulse train serves to synchronize the logic system; this is analogous to the sync pulses in a tv signal. The output from the clock multivibrator is applied to a flip-flop (bistable multivibrator), which in turn energizes another flip-flop and four AND gates.

DIGITAL-LOGIC WAVEFORM RELATIONSHIPS

Digitial-logic equipment is primarily a network of *gates* and *flip-flops*, and it will be shown later that a flip-flop is an arrangement of cross-connected gate ICs. Digital-logic equipment operation is *synchronized* by pulses (clock pulses) produced by a clock generator. Clock generators operate at various speeds, up to many megahertz. Therefore, oscilloscopes used in troubleshooting digital-logic equipment may require an unusually high vertical-amplifier frequency response. For example, a typical gate can be operated by 30-MHz square waves. However, when several gates are connected into a network, the operating speed of the system is necessarily less than that of a single gate. The most basic forms of logic gates are the AND gate, the NAND gate, the OR gate, and the NOR gate, as shown in Fig. 11-2. Each gate has two inputs in these examples; but, we will encounter

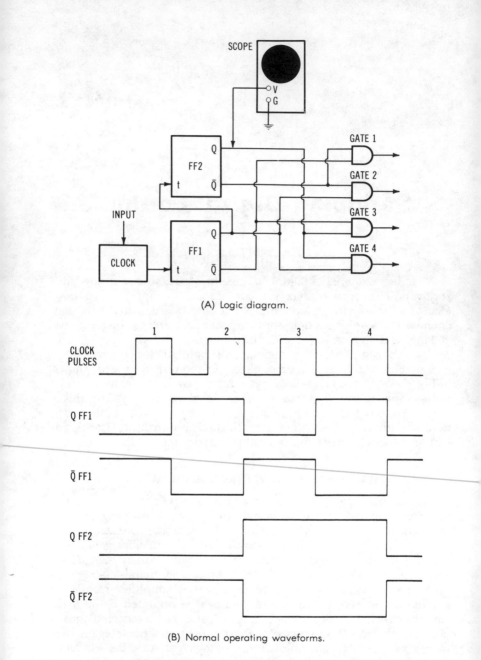

(A) Logic diagram.

(B) Normal operating waveforms.

Fig. 11-1. A simplified block diagram for a 4-channel digital-logic section in a scanner-monitor radio receiver.

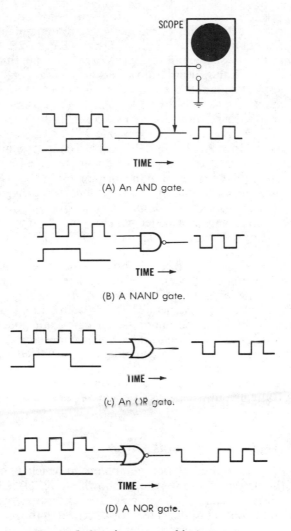

SCOPE

TIME →

(A) An AND gate.

TIME →

(B) A NAND gate.

TIME →

(c) An OR gate.

TIME →

(D) A NOR gate.

Fig. 11-2. Signal responses of logic gates.

gates that have three, four, or more inputs. However, the number of inputs to a gate does not change its basic mode of response.

Observe in Fig. 11-2 that a square-wave input and a rectangular-wave input are applied to the AND gate. In turn, two square waves normally appear at the output. Thus, *an AND gate produces an output only when both of its inputs are driven logic-high.* In other words, during the time that either of the inputs to the AND gate is zero (logic low), there is a zero output from the AND gate. However, during the time that both of the inputs of the AND gate are logic-high, there is a logic-high output from the AND gate. Note next (Fig. 11-2B) that a NAND gate operates in the same manner as an AND gate, except that its output is inverted. That is, the output from a NAND gate is logic-high during the time that both of its inputs are logic-low. However, when both of the inputs of the NAND gate are driven logic-high, the output of the NAND gate will go logic-low. Suppose that a NAND gate is provided with three inputs; then, the output of the NAND gate will be logic-high until all three inputs are driven logic-high. With all three inputs simultaneously logic-high, the output of the NAND gate will go logic-low.

Next, observe the operation of the OR gate in Fig. 11-2C. Note that *an OR gate produces an output if either of its inputs is driven logic-high; the OR gate also produces an output if both of its inputs are driven logic-high.* If an OR gate is provided with four inputs, it will produce an output when one or more of the inputs are driven logic-high. Note next that a NOR gate (Fig. 11-2D) operates in the same manner as an OR gate except that its output is inverted. In other words, the output terminal of a NOR gate is logic-high as long as both of its inputs are logic-low. However, if one or both of the inputs of the NOR gate is driven logic-high, its output goes logic-low. Then, when both of the inputs go logic-low once more, the output of the NOR gate again goes logic-high. Although single-trace and dual-trace oscilloscopes are very useful in troubleshooting digital-logic equipment, specialized digital-logic scopes may have up to 16 channels. An example of the use of a 3-channel scope in checking the operation of an AND gate is shown in Fig. 11-3. Basically, the scope displays a *timing diagram*, as depicted in Fig. 11-4.

FAULT PATTERNS

Consider a unit of digital-logic equipment in which an open circuit occurs at the output of an AND gate, as shown in Fig. 11-5. Test signals are applied at the inputs of AND gate U1, and an output signal is displayed at the output of the gate. However, when the scope probe is moved ahead in the same circuit, there is only a dc "bad level" output. In TTL and DTL logic circuitry, the range from 1.4 to 1.5 volts indicates a fault (in this example, an open circuit). Note that a "bad level" will be interpreted by the following circuits as a logical high level. A logical high level is at least 2.4 volts and a logical low level is 0.4 volt or less, as depicted in Fig. 11-6. The output waveform from a gate must rise and fall within a sufficiently short time so that subsequent gates in the system can be operated properly. In turn, excessive rise time is another basic fault condition, as exemplified in Fig. 11-7. Note that the rise time of a digital signal will always be

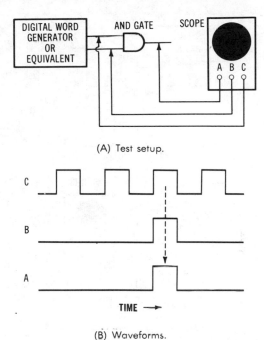

(A) Test setup.

(B) Waveforms.

Fig. 11-3. Checking the operation of an AND gate using a 3-channel scope.

slowed more or less by lines (interconnects) that are installed for the operation of peripheral memories, or other logic units. Therefore, the clock speed may need to be slowed if interconnects impose an excessive rise time.

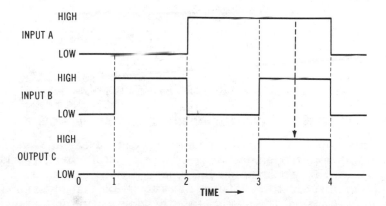

Fig. 11-4. A timing diagram for an AND gate.

173

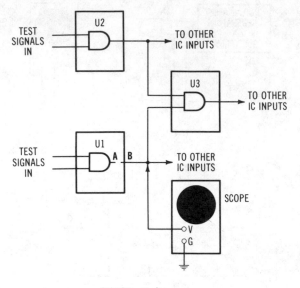

(A) Circuit diagram.

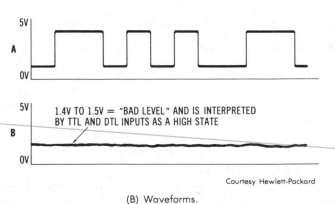

5V

A

0V

5V — 1.4V TO 1.5V = "BAD LEVEL" AND IS INTERPRETED
BY TTL AND DTL INPUTS AS A HIGH STATE

B

0V

(B) Waveforms.

Fig. 11-5. Example of a "bad level" output due to an open circuit.

OPERATING WAVEFORMS IN SIMPLE DIGITAL NETWORKS

A basic clock oscillator (multivibrator) circuit is exemplified in Fig. 11-8, with normal operating waveforms. In a typical system, V_{cc} may be 6 volts. Faulty operation in this type of circuitry can result either from defective capacitors or failing transistors. Note that power failure occurs on occasion. This may be caused by trouble in the power supply itself, in the power distribution cabling, or in the power decoupling

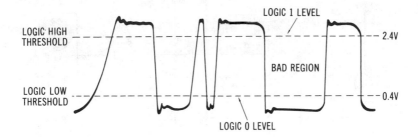

Fig. 11-6. Logic-high and logic-low thresholds in TTL and DTL circuitry.

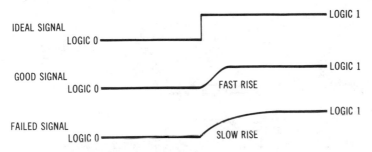

Fig. 11-7. Excessive rise time is a fault condition.

circuitry included on each circuit board. Another common cause of trouble symptoms is a connecting wire that has become open or shorted to ground. Transistors tend to fail catastrophically. Short circuits may overheat and damage resistors in the associated circuit.

Observe the Exclusive-OR (XOR) gate shown in Fig. 11-9. An XOR gate can be formed from two AND gates, an OR gate, and two inverters (polarity inverters). A polarity inverter, such as a common-emitter stage, inverts the input signal; a positive-going signal is outputted as a negative-going signal, and vice versa. Thus, a logic-high input signal is outputted as a logic-low signal by an inverter. An XOR gate functions to produce a logic-high output when one of its inputs is logic-high, and its other input is logic-low. On the other hand, the output of an XOR gate remains logic-low when both of its inputs are logic-low, or when both of its inputs are logic-high. In other words, *an XOR gate produces an output only when its inputs are at opposite logic levels; an XOR gate produces zero output when its inputs are at the same logic level.* In turn, an XOR gate responds to input pulses as shown in Fig. 11-9D. An XOR gate connected with an AND gate is called a *half-adder* because its output is the sum of 0 and 1, of 1 and 0, or of 1 and 1 (Fig. 11-10).

A *full adder* consists of two half-adders and an OR gate, as shown in Fig. 11-11. A full adder provides for a carry-in (C') as well as for a

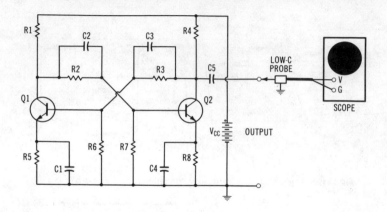

(A) Circuit diagram.

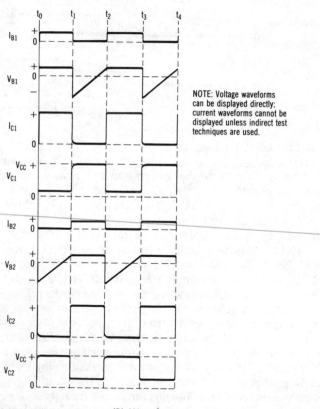

NOTE: Voltage waveforms can be displayed directly; current waveforms cannot be displayed unless indirect test techniques are used.

(B) Waveforms.

Fig. 11-8. A basic clock oscillator circuit and its normal waveforms.

176

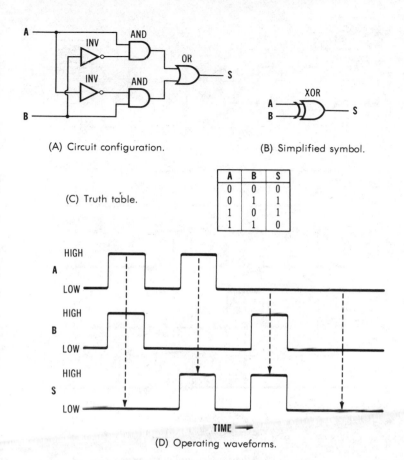

(A) Circuit configuration.

(B) Simplified symbol.

(C) Truth table.

A	B	S
0	0	0
0	1	1
1	0	1
1	1	0

(D) Operating waveforms.

Fig. 11-9. An Exclusive-OR gate, showing its individual gates, inverters, waveforms, and truth table.

carry-out (C) terminal. The provision of a carry in line permits the connection of two full adders in series, for the simultaneous (parallel) addition of large binary numbers. In addition to checking the timing diagram for a unit of digital-logic equipment, the pulse widths and the elapsed time between pulses can be measured with a triggered-sweep scope that has a calibrated time base. Note that if an "intelligent oscilloscope" is employed, the operator can place intensity markers at the start and stop intervals to be timed (t_1 and t_2 in Fig. 11-12). Then, the LED readout, in the example given in Fig. 11-12, automatically and continuously indicates the time between the two markers (1.92×10^{-6} second in our example). This feature provides operator convenience, inasmuch as the calculation of time is performed automatically.

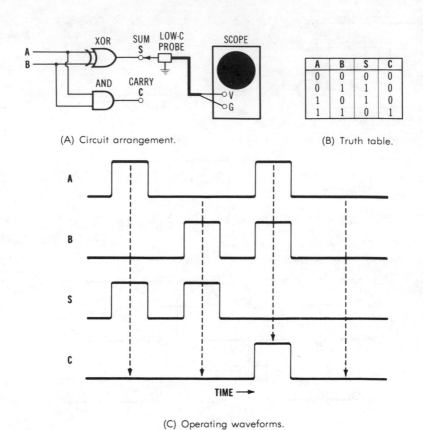

(A) Circuit arrangement.

(B) Truth table.

A	B	S	C
0	0	0	0
0	1	1	0
1	0	1	0
1	1	0	1

(C) Operating waveforms.

Fig. 11-10. A half-adder configuration.

BINARY COUNTERS

Binary counters are widely used in digital equipment. Many counters employ J-K flip-flops, as shown in Fig. 11-13. A J-K flip-flop has two conditioning inputs (J and K) and one clock input (T). If both conditioning inputs are disabled prior to a clock pulse, the flip-flop does not change condition when a clock pulse occurs. If the J input is enabled, and the K input is disabled, the flip-flop will assume a 1 condition (Q output logic-high; $\overline{Q}$ output logic-low) upon arrival of a clock pulse. Note that $\overline{Q}$ denotes NOT Q, or the inverse (complement) of Q. Next, if the K input is enabled, and the J input is disabled, the flip-flop will assume the 0 condition when a clock pulse arrives. If both the J and K inputs are enabled prior to the arrival of a clock pulse, the flip-flop will complement (or assume the opposite state) when the

178

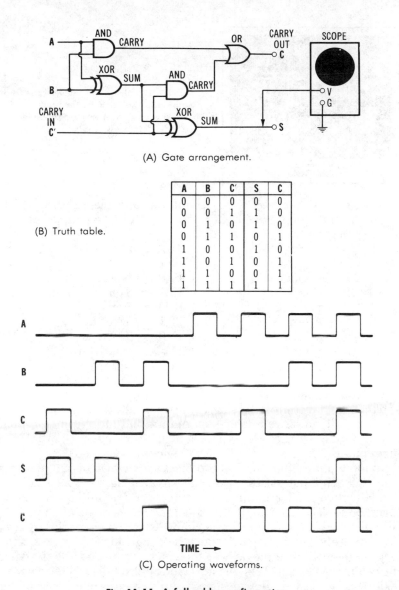

(A) Gate arrangement.

(B) Truth table.

A	B	C′	S	C
0	0	0	0	0
0	0	1	1	0
0	1	0	1	0
0	1	1	0	1
1	0	0	1	0
1	0	1	0	1
1	1	0	0	1
1	1	1	1	1

(C) Operating waveforms.

Fig. 11-11. A full adder configuration.

clock pulse occurs. In other words, at the application of a clock pulse, a 1 on the J input sets the flip-flop to the 1 or "on" state; a 1 on the K input resets the flip-flop to its 0 or "off" state. If a 1 is simultaneously applied to both inputs, the flip-flop will change state, regardless of its previous state.

179

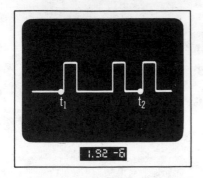

Fig. 11-12. Using an oscilloscope that automatically indicates the elapsed time between two operator-positioned intensity markers.

Next, observe the 4-bit binary counter arrangement shown in Fig. 11-14. A bit denotes a *binary digit*, such as 1 or 0. Examples of 4-bit binary numbers are:

0000 = Zero	1000 = Eight
0001 = One	1001 = Nine
0010 = Two	1010 = Ten
0011 = Three	1011 = Eleven
0100 = Four	1100 = Twelve
0101 = Five	1101 = Thirteen
0110 = Six	1110 = Fourteen
0111 = Seven	1111 = Fifteen

The counter depicted in Fig. 11-14 counts from 0 (0000) to 15 (1111), and then resets itself to 0. As seen in the timing diagram of Fig. 11-14B, flip-flop FF1 produces a logic-high output pulse for each two input pulses; flip-flop FF2 produces a logic-high output pulse for every four input pulses; flip-flop FF3 produces a logic-high output pulse for every eight input pulses; flip-flop FF4 produces a logic-high output pulse for every sixteen input pulses. Thus, this circuit diagram is also called a divide-by-two configuration. Observe that the J and K inputs are tied together on each flip-flop; this is called the *toggle* mode of operation. In other words, flip-flop FF1 will change state on every clock pulse, provided only that the Logic-1 input is held logic-high. Of course, if the Logic-1 input is driven logic-low, the counter stops operation. After the count reaches 15 (1111), the next clock pulse will reset the chain to 0000 because each flip-flop must necessarily change state from 1 to 0 at this time. The AND gates are included in this counter design simply to obtain a faster response (minimum propagation delay).

Another term for the 4-bit binary-counter arrangement shown in Fig. 11-14 is a *synchronous binary counter*. Note that if the AND gates are omitted, the flip-flop chain is then called a *ripple-carry counter*. A ripple-carry counter requires a certain amount of time to change state.

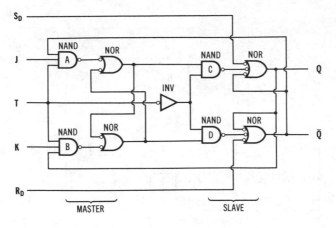

(A) Gate arrangement.

(B) Logic symbol.

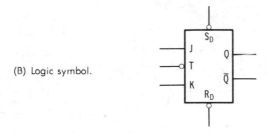

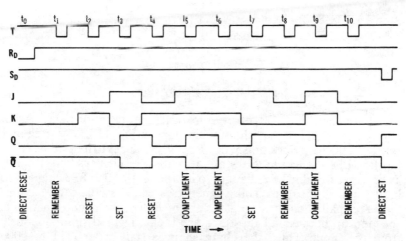

(C) Waveform relations.

Fig. 11-13. Operation of a J-K flip-flop.

181

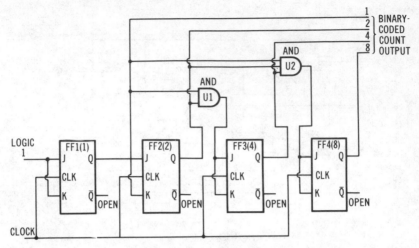

NOTE: Binary weights of the flip-flops are shown in parentheses.

(A) Circuit arrangement.

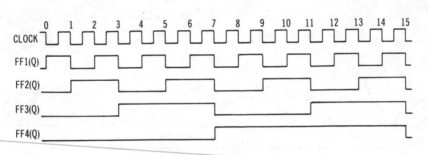

(B) Operating waveforms.

Fig. 11-14. A 4-bit binary counter.

This state-change time involves more or less propagation delay. As an illustration, suppose that all four flip-flops in Fig. 11-14 are in their 1 state. On the next clock pulse, flip-flop FF1 changes to its 0 state. After flip-flop FF1 changes to its 0 state, flip-flop FF2 changes to its 0 state. Then, after flip-flop FF2 changes to its 0 state, flip-flop FF3 changes to its 0 state, and so on. This sequence is called ripple-carry action. Observe that when the AND gates are included, the counter action is speeded up because all flip-flops are then triggered simultaneously from the input. Thus, the chain changes state from 1111 to 0000 simultaneously instead of sequentially, as when the synchronous configuration is used. In turn, the clock can be operated at a considerably higher frequency.

DIGITAL PULSE WAVESHAPES

In theory, digital pulses have square corners; the leading edge ideally has zero rise time, and the trailing edge has zero fall time. However, in practice, substantial departures from the ideal often occur. In one type of digital equipment, the pulses have an appearance as shown in Fig. 11-15. The tops of the pulses are rounded, and the rise time is greater than zero; the fall time is longer than the rise time. Thus, the pulses are somewhat spike-shaped, instead of being precisely rectangular. However, this is not a matter for practical concern, as long as the distortion is not so serious that the gates malfunction. Tolerable waveshape distortion is chiefly a matter for experienced judgment; it is difficult to set down hard-and-fast rules in this regard. *Propagation delay* was noted above. While not a form of pulse distortion per se, excessive propagation delay in a digital system inevitably leads to unreliable operation. Therefore, in troubleshooting a digital system with the oscilloscope, the technician should be alert for abnormal propagation delays, as well as for excessive pulse-shape distortion.

Excessive propagation delay and waveshape distortion can give rise to *glitches* in otherwise well-designed digital equipment. Glitches are spurious pulses produced by marginal gate operation. Some examples of glitches are given in Fig. 11-16. It is evident that glitches will cause equipment malfunction, inasmuch as a gate or a flip-flop cannot "tell the difference" between a glitch and a data pulse or waveform. Glitches can also occur because of power-supply switching transients. Sometimes glitches are induced into digital equipment by strong external fields, as in industrial installations. In any case, the oscilloscope is the most useful troubleshooting instrument for analyzing and tracking down the source of glitches.

Note also that glitches are occasionally very narrow and may have a very fast rise and fall time. In such a case, a high-performance scope is

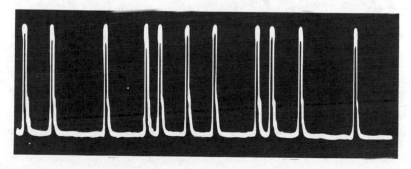

Fig. 11-15. Digital pulses may not have ideal waveshapes.

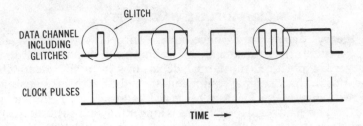

Fig. 11-16. Example of "glitches" in a digital data signal.

required to display a visible glitch pattern. A considerably higher writing speed may be required than for displaying the normal operating waveforms in the digital equipment. Troubleshooting procedures are sometimes impeded by the transient occurrence of a glitch. In other words, if the technician knows that a trouble symptom is being caused by a glitch, but the glitch occurs randomly at widely separated intervals, it may be a time-consuming task to identify the trouble area and to close in on the defective device or component. In such a case, it is often helpful to use a storage-type scope as a monitor. Thereby, the technician can look at the screen from time to time to determine whether a glitch has been displayed. It releases the scope operator from the necessity of closely and continuously watching the screen to "catch the glitch on the fly." In practice, glitches as narrow as 25 nanoseconds are encountered; only the most sophisticated oscilloscopes are capable of "catching" and displaying such narrow pulses.

With reference to Fig. 11-17, the exemplified rounding of the output pulse from the MOS ratio-inverter device is normal and should not be a matter for concern by the troubleshooter. Of course, excessive corner rounding, accompanied by an abnormally long rise time, can cause equipment malfunction. Permissible tolerances on waveshapes are largely learned by experience. For example, the distorted waveforms across C1 and C2 in Fig. 11-18 might well be regarded with alarm by the

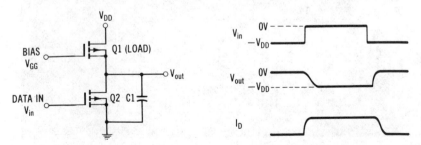

Fig. 11-17. Normal rounding of digital pulses by an MOS ratio-inverter device.

184

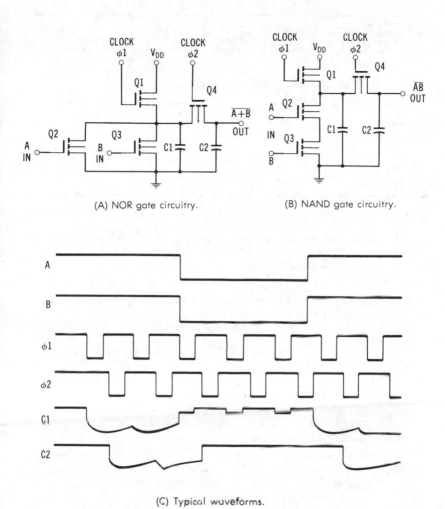

(A) NOR gate circuitry.

(B) NAND gate circuitry.

(C) Typical waveforms.

Fig. 11-18. Basic dynamic 2-phase gates.

apprentice technician. However, these waveforms are entirely normal for these MOS devices, as may be determined by consulting the manufacturer's technical data. However, normal waveforms are less than ideal in TTL circuitry. For example, the 4-channel display exemplified in Fig. 11-19 for the Q outputs of a TTL decade divider will show some departures from the ideal waveshapes. Thus, the clock pulses will exhibit tilt along their bottom excursions; the next lower waveform will have a visible tilt along its top excursions. All of the pulses will show some evidence of overshoot or undershoot.

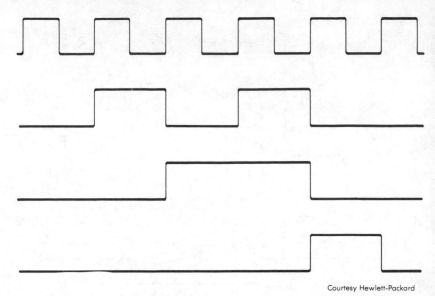

Fig. 11-19. Normal timing diagrams for the Q outputs of a TTL decade divider.

DATA-DOMAIN DISPLAYS

With the advent of the digital revolution, specialized high-technology oscilloscopes, such as the one shown in Fig. 11-20, tend to dominate the lab-type category of designer's and troubleshooter's display instruments. A digital data-field display is distinguished from a conventional analog (time/frequency) waveform display by the fact that the former shows sequences of digital events as 1s and 0s with respect to computer clock time, whereas, the latter shows the variation of an electrical quantity with real time (the actual time in which the physical events occurred).

Some digital-troubleshooting oscilloscopes provide a choice of timing-diagram or data-domain displays, as shown in the correlation example of Fig. 11-21. Thus, the troubleshooter may switch his 6-channel scope either to display timing diagrams or to display data fields. In a preliminary analysis of equipment malfunction, it is often quite helpful to observe the data-domain display of binary digits. This display can be directly compared with truth tables, and requires no "interpretation" from wave trains into digital "words." In turn, the channel that exhibits a malfunction is directly identifiable. On the other hand, analysis of the malfunction can generally be made to better advantage on the basis of a timing-diagram waveform display. For example, if the malfunction happens to be caused by a glitch, the spurious pulse can be tracked down more confidently on the basis of a conventional waveform display.

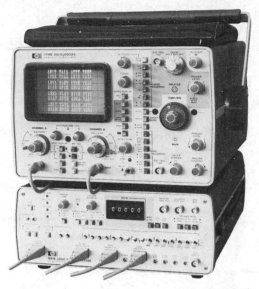

Fig. 11-20. A high-technology oscilloscope for use in digital logic-state analysis.

GLITCHES CAUSED BY "RACES"

A *race* condition is present in a digital circuit when two or more pulse inputs that are applied to the circuit arrive simultaneously. It is apparent that if the *order* in which the *two pulses are applied* to a device

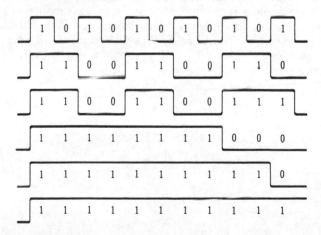

Fig. 11-21. A representative correlation of timing-diagram and data-domain displays.

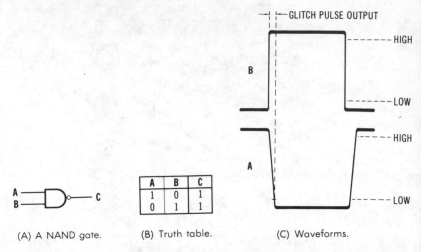

(A) A NAND gate.

A	B	C
1	0	1
0	1	1

(B) Truth table.

(C) Waveforms.

Fig. 11-22. A "race" condition occurs if input B goes logic-high before input A attains the logic-low level.

determine its output state, a critical *race condition* exists. Refer to Fig. 11-22. This is a two-input NAND gate which can malfunction if input B goes logic-high *before* input A attains the logic-low level. In this situation, inputs A and B are both driven logic-high momentarily, with the result that a spurious pulse is produced at output C. Of course, the spurious pulse ends as soon as input A reaches the logic-low threshold. This negative pulse is a typical glitch. In digital-equipment design, propagation delays inherent in the A and B input circuits may be "tailored" to avoid a race condition. However, avoidance of glitches then depends on permissible waveform tolerances. Thus, in the event that the pulse sent to input A develops an excessive fall time, a race condition will occur, with the development of a glitch output from the NAND gate.

Index